Analysis of Chemical Residues in Agriculture

Analysis of Chemical Residues in Agriculture

SÍLVIO VAZ, JR.

Senior Scientist, Brazilian Agricultural Research Corporation (Embrapa), Brasília, DF, Brazil; Professor of Environmental Engineering, Federal University of Ouro Preto, Ouro Preto, Brazil

Elsevier
Radarweg 29, PO Box 211, 1000 AE Amsterdam, Netherlands
The Boulevard, Langford Lane, Kidlington, Oxford OX5 1GB, United Kingdom
50 Hampshire Street, 5th Floor, Cambridge, MA 02139, United States

Copyright © 2021 Elsevier Inc. All rights reserved.

No part of this publication may be reproduced or transmitted in any form or by any means, electronic or mechanical, including photocopying, recording, or any information storage and retrieval system, without permission in writing from the publisher. Details on how to seek permission, further information about the Publisher's permissions policies and our arrangements with organizations such as the Copyright Clearance Center and the Copyright Licensing Agency, can be found at our website: www.elsevier.com/permissions.

This book and the individual contributions contained in it are protected under copyright by the Publisher (other than as may be noted herein).
Notices
Knowledge and best practice in this field are constantly changing. As new research and experience broaden our understanding, changes in research methods, professional practices, or medical treatment may become necessary.

Practitioners and researchers must always rely on their own experience and knowledge in evaluating and using any information, methods, compounds, or experiments described herein. In using such information or methods they should be mindful of their own safety and the safety of others, including parties for whom they have a professional responsibility.

To the fullest extent of the law, neither the Publisher nor the authors, contributors, or editors, assume any liability for any injury and/or damage to persons or property as a matter of products liability, negligence or otherwise, or from any use or operation of any methods, products, instructions, or ideas contained in the material herein.

British Library Cataloguing-in-Publication Data
A catalogue record for this book is available from the British Library

Library of Congress Cataloging-in-Publication Data
A catalog record for this book is available from the Library of Congress

ISBN: 978-0-323-85208-1

For Information on all Elsevier publications
visit our website at https://www.elsevier.com/books-and-journals

Publisher: Megan R. Ball
Acquisitions Editor: Nancy Maragioglio
Editorial Project Manager: Ruby Smith
Production Project Manager: Bharatwaj Varatharajan
Cover Designer: Greg Harris

Typeset by MPS Limited, Chennai, India

Dedication

I would like to dedicate this book to my father, Silvio Vaz, who is a farmer in Brazil since his childhood.

Contents

Preface — xiii

1. Introduction to organic and inorganic residues in agriculture — 1
- 1.1 Definition of agricultural residues and chemical residues — 5
- 1.2 Main organic chemical residues — 5
- 1.3 Main inorganic chemical residues — 7
- 1.4 Degradation processes — 9
- 1.5 Metabolites — 11
- 1.6 International control policies for residues in crops and food — 12
- 1.7 Food safety and standardization — 13
- 1.8 New trends in agrochemistry — 16
- 1.9 Conclusion — 17
- References — 18

2. Agricultural matrices — 21
- 2.1 Environmental matrices—soil and water — 22
  - 2.1.1 Soil — 23
  - 2.1.2 Water — 25
- 2.2 Food and feed — 27
- 2.3 Meat and other animal products — 30
- 2.4 Beverages and fruit juices — 33
- 2.5 Agroindustrial waste — 33
  - 2.5.1 Pig slurry — 36
  - 2.5.2 Waste of beer fermentation — 36
- 2.6 Conclusion — 36
- References — 37

3. Toxicology in agriculture — 39
- 3.1 Definition of main terms in ecotoxicology and toxicology — 39
  - 3.1.1 Ecotoxicology — 39
  - 3.1.2 Toxicology — 55
- 3.2 Disorders and diseases associated to agrochemical residues — 78
- 3.3 Dietary exposure — 78
- 3.4 Occupational exposure — 79
- 3.5 Environmental exposure — 80

	3.6	Conclusion	83
	References		83

4. Fundamentals of analytical chemistry   85

	4.1	Analytical chemistry in the 21st century	86
	4.2	Figures of merit	86
		4.2.1 Accuracy	87
		4.2.2 Linearity	87
		4.2.3 Limit of detection and limit of quantification	87
		4.2.4 Precision	88
		4.2.5 Sensitivity or sensibility	89
		4.2.6 Selectivity	89
		4.2.7 Robustness	89
		4.2.8 Recovery	90
	4.3	Developing an analytical method	90
		4.3.1 Calibration	92
	4.4	Validating an analytical method	94
		4.4.1 Interlaboratory studies	96
		4.4.2 Interlaboratory comparisons	97
		4.4.3 Systematic evaluation of factors influencing results	97
		4.4.4 Evaluation of uncertainty of results generated	97
		4.4.5 Repeatability and reproducibility	98
		4.4.6 Accreditation of an analytical laboratory for residue analysis	99
	4.5	Chemometrics	100
	4.6	Quality control and quality assurance	103
	4.7	Green analytical chemistry	105
	4.8	Conclusion	107
	References		107

5. Main analytical techniques   111

	5.1	Spectroscopic techniques	114
		5.1.1 Absorption of UV–vis radiation, or molecular spectrophotometry	114
		5.1.2 Emission of UV–vis radiation, or fluorescence	117
		5.1.3 Infrared molecular spectroscopy	119
		5.1.4 Atomic absorption spectrometry	122
		5.1.5 Atomic emission spectrometry or optical emission spectrometry	123
		5.1.6 X-ray emission spectrometry	125
		5.1.7 Nuclear magnetic resonance	126
	5.2	Mass spectrometry	129

5.3		Chromatographic techniques	131
	5.3.1	Gas chromatography	132
	5.3.2	Liquid chromatography	134
5.4		Electrochemical techniques	138
	5.4.1	Potentiometry	138
	5.4.2	Voltammetry	138
	5.4.3	Electrophoresis	139
5.5		Probes and sensors	140
5.6		Bioassays	143
5.7		Sampling	146
	5.7.1	Environmental samples	146
	5.7.2	Food samples	149
5.8		Sample preparation	150
5.9		Conclusion	167
References			168

6. Analytical methods to selected matrices — 171

6.1		Analytical method for organic residues (pesticides) in the environment	171
	6.1.1	Method	172
	6.1.2	Source	172
	6.1.3	Scope and application	172
	6.1.4	Description	172
6.2		Analytical method for organic residues (pesticides) in vegetables and fruits, and their wastes	174
	6.2.1	Method	174
	6.2.2	Source	174
	6.2.3	Scope and application	175
	6.2.4	Description	175
6.3		Analytical method for organic residue (veterinary drug) in meat	177
	6.3.1	Method	177
	6.3.2	Source	177
	6.3.3	Scope and application	177
	6.3.4	Description	178
6.4		Analytical method for organic residues (additives) in processed product (food, beverage, and feed)	179
	6.4.1	Method	180
	6.4.2	Source	180
	6.4.3	Scope and application	180
	6.4.4	Description	180

6.5	Analytical method for inorganic residues in the environment	182
	6.5.1	Method	182
	6.5.2	Source	183
	6.5.3	Scope and application	183
	6.5.4	Description	185
6.6	Analytical method for inorganic residue in juice	186
	6.6.1	Method	187
	6.6.2	Source	187
	6.6.3	Scope and application	187
	6.6.4	Description	188
6.7	Analytical method for inorganic residue in meat	191
	6.7.1	Method	191
	6.7.2	Source	191
	6.7.3	Scope and application	191
	6.7.4	Description	191
6.8	Conclusion	193
References	194

7. Analytical chemistry towards a sustainable agrochemistry — 195

7.1	What is sustainable agrochemistry?	195
	7.1.1	The demand for sustainability	197
	7.1.2	The agrochemical classes	199
	7.1.3	Undesirable effects from agrochemical usages	201
	7.1.4	Pillars of sustainability	201
	7.1.5	Agrochemical regulation and commercialization	202
7.2	Outlook of emerging analytical technologies	205
	7.2.1	Artificial intelligence in analytical chemistry	207
7.3	Outlook of innovative analytical approaches	210
	7.3.1	Combined automation and miniaturization approach for surfactant presence	211
	7.3.2	Combined automation and miniaturization approach for multiresidue presence	212
7.4	Conclusion	213
References	213

8. Sorption study for environmental purpose — 217

8.1	Sorption isotherms	217
8.2	Methodology	221
	8.2.1	Soil characterization	221
	8.2.2	Sorption soils-agrochemical	221

	8.3	Results and discussion	226
		8.3.1 Soil characterization	226
		8.3.2 Sorption soil-agrochemical	226
	8.4	Conclusion	234
	References		235

9. A practical guide for residue analysis — 237

	9.1	Establishing (indeed) a quality control	237
	9.2	Avoiding problems in the sampling step	238
		9.2.1 Chain of custody	239
	9.3	Laboratory management	240
	9.4	Reporting and interpreting analytical results	240
	9.5	Estimation of measurement uncertainty of results	241
	9.6	Conclusions	242
	References		243

10. General remarks and conclusions — 245

	10.1	Remarks	245
	10.2	Conclusion	250
	References		253

Index — *255*

Preface

Agriculture remains one of the most strategic sectors for the global economy and its well-being. However, it is seen as a source of environmental and health concerns due, mainly, to the high amount of agrochemicals, such as pesticides and fertilizers, used in production systems around the world; moreover, when we consider the livestock production systems they also applied large amounts of veterinary drugs to treat illness and to promote an increase in productivity.

Unfortunately, agrochemicals generate residues, which are present in crops, fruits, meats, and processed products (food and feed) that need permanent monitoring and control. In this way, the measure of chemical residues in agriculture and livestock is paramount, and analytical chemistry can certainly contribute to the generation of wealth and health for modern society. The use of advanced analytical techniques such as chromatography, spectroscopy, spectrometry, electrochemistry, among others, can generate reliable information about the quality of several products and raw materials. Furthermore, the current demand from modern society for a sustainable production of food has promoted the development of policies considering aspects such as reducing negative impacts on the environment and overall health, friendly materials and molecules, bioactive compounds, etc.

This book's goal is to present a large variety of analytical technologies and methods in order to help professionals, researchers, and graduate and undergraduate students in the understanding and application of the most adequate analytical approach for the detection and quantification of chemical residues—organic and inorganic—in several agricultural matrices, as crops, fruits, meats, food, feed, soil, and water. Furthermore, this book aims to demonstrate the importance of analytical chemistry in the contribution toward the environment's health, through the application of conventional and innovative techniques and methods of analysis with a positive and direct influence on the reduction of negative impacts from agricultural systems on society. These statements are based on large efforts seeking the technical and scientific development of analytical chemistry, which have been taking place over decades of work by the chemical community.

The author carried out a careful survey of the main analytical techniques and instrumentation in use, using his academic and professional experiences. Moreover, the author also considered trends observed in current analytical sciences, regulatory legislation, and other relevant aspects, such as the principles of green chemistry and the proposal of a more sustainable analytical chemistry in order to reach sustainable agricultural chains.

Ten chapters compose the book: Chapter 1, Introduction to Organic and Inorganic Residues in Agriculture, deals with the definition and examples of the most common organic and inorganic residues generated by agriculture, livestock, and their processing, which are a subject of analyses according to established international values (e.g., pesticides, fertilizers, veterinary drugs, etc.). Furthermore, aspects of food supply chain, degradation processes, metabolites from these residues, and new trends in agrochemistry are considered.

Chapter 2, Agricultural Matrices, deals with the chemical composition and physicochemical properties of the main agricultural matrices, for example, crops, food, feed, meats and other animal products, soil, water, and agroindustrial waste. Moreover, aspects of the interaction and fate of residues and these matrices are discussed in order to understand the relevance of chemical analyses to determine chemical residues in these matrixes.

Chapter 3, Toxicology in Agriculture, deals with an introduction to the basis of toxicology and ecotoxicology to be applied to agricultural systems, products, and their consumption. Moreover, aspects related to dietary, occupational, and environmental exposures and associated disorders and diseases are explored in order to reinforce the relevance of chemical analyzes.

Chapter 4, Fundamentals of Analytical Chemistry, deals with the main fundamentals of analytical chemistry to be applied in the agrochemical residue analyses in order to reach the reliability of the analytical results. A solid mathematical basis is presented, highlighting those statistics related to the analytical method development and validating. Moreover, aspects of chemometrics, green analytical chemistry and quality assurance/quality control are presented and discussed.

Chapter 5, Main Analytical Techniques, deals with the most common analytical techniques applied in the analyses of agrochemical residues. Physical and chemical principles, block diagrams, and application examples are depicted and discussed. Moreover, recent developments in analytical technologies are considered also.

Chapter 6, Analytical Methods to Selected Matrices, deals with methodologies to determine several agrochemical residues in selected matrices related to food and the environment, subjects to control and monitoring. Its content is based on the most recent scientific literature with a critical evaluation for each method. The general objective is to offer a technical basis in order to choose the best methodology to be applied to real situations, providing information to the analyst for a certain method development and/or its adaptation.

Chapter 7, Analytical Chemistry Towards a Sustainable Agrochemistry, deals with the concept of sustainable agrochemistry and its applicability in modern agriculture in order to demonstrate the contribution of the modern analytical chemistry to reach these statements. Emerging analytical techniques, especially nanosensors are presented and discussed. A set of innovative analytical approaches are also exploited.

Chapter 8, Sorption Study for Environmental Purpose, deals with the construction and interpretation of sorption isotherms for Brazilian tropical soils and the veterinary antibiotic oxytetracycline and the involved molecular interactions, in order to understand the fate of the antibiotic residue in the environment, which is subject to evaluation for regulation. Then a proposed experimental strategy was applied and validated using HPLC-UV-Vis and mathematical models proving that the sorption study can supply data to predict the deleterious effect of a certain agrochemical molecule.

Chapter 9, A Practical Guide for Residue Analysis, deals with relevant difficulties to be overcome in order to reach the best conditions for the residue analysis. These issues comprise of difficulties for the establishment of a feasible quality control strategy, problems in the sampling step, laboratory management, a correct reporting and interpreting analytical results, and the correct estimation of measurement uncertainty of results.

Finally, Chapter 10, General Remarks and Conclusions, deals with the more relevant information described by the previous chapters in order to highlight the practical application of them—remarks for statements and conclusions for critical summarizing. Furthermore, it will help to link the subject of each chapter through a holistic view addressed to solve representative issues related to modern agriculture by means of modern analytical chemistry.

Good lecture!
Sílvio Vaz Jr.

CHAPTER 1

Introduction to organic and inorganic residues in agriculture

Agriculture is a large economic activity that comprises the production of raw material for food, feed, chemicals, pharmaceuticals, among other products. According the Food and Agriculture Organization of the United Nations (2018), the global agricultural production in 2016 achieved 4.7 thousand tons for the five main items produced (sugarcane, maize, wheat, rice, potatoes). For the livestock production, it achieved 27.6 thousand heads for the five main items (chickens, cattle, ducks, sheep, and goats). From this point of view, agricultural practices and food chains need chemical substances such as:
- pesticides
- fertilizers
- additives
- stabilizers
- preservatives
- antioxidants
- antibiotics
- sanitizers
- others

These chemical substances are known as agrochemicals and are used from tillage to harvest and processing steps. The production steps and uses generate a variety of sources of chemical residues that demand attention due to their negative impact on the environment and overall human health.

We can observe in Fig. 1.1 an example of supply food chain with the sources of agrochemicals entrance that can generate chemical residues in food and their wastes. It can be directly related to the negative impact on the environment and human health.

The five entrances of chemicals described in Fig. 1.1 can be understood as:
- Entrance 1: pesticides and fertilizers (mainly) are applied to promote the best conditions for the agricultural production.
- Entrance 2: additives and stabilizers are used to guarantee physicochemical, nutritional and organoleptic properties for the final product.

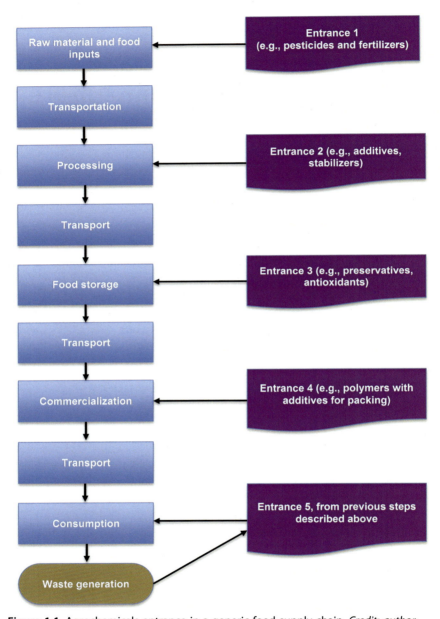

Figure 1.1 Agrochemicals entrance in a generic food supply chain. *Credit: author.*

- Entrance 3: preservatives and antioxidants are used to maintain those properties of raw material and its processing without any degradation process.
- Entrance 4: packing materials are used to maintain the quality of those products under suitable conditions for their commercialization in ambient as a supermarket.
- Entrance 5: all those chemicals from entrance 1 to 4 can be concentrated during the consumption step and the final disposition of the agricultural production, with a possible residues accumulation.

As examples of entrance, in 2016 the world total agricultural use of chemical or mineral fertilizers was 110 Mt nitrogen (N), 49 Mt phosphate (P_2O_5), and 39 Mt potash (K) (Food and Agriculture Organization of the United Nations, 2018). Regarding also Food and Agriculture Organization (FAO), for pesticides the main agricultural producers—the United States, Brazil, European Union, Russia and China—had a use in the same year up to 1.8 million tons of active ingredients for each country.

A certain agriculture production chain can involve, also, several agrochemicals and other chemicals application, as depicted in Fig. 1.2. Again,

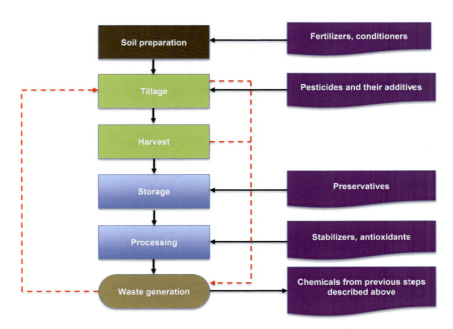

Figure 1.2 Description of a generic agricultural system. *Credit: author.*

it can be directly related to the negative impact on the environment and human health.

It is possible to observe in Fig. 1.2 some similarities regarding to the chemical entrance with elements of Fig. 1.1. However, the cyclic use of the biomass residues from harvest and production describes a particularity of the more sustainable agricultural systems.

To guarantee the confidence and quality of agricultural products and those new products derived from them, is paramount the use and application of analytical techniques and methods to determine the presence or absence of those chemicals highlighted in Figs. 1.1 and 1.2. Fig. 1.3 depicts the fate of chemicals in agrifood chains with the application of chemical analysis to determine the presence or absence of them—as residues—in products (e.g., food) and in the environment.

A very common analytical technique used in this approach is the chromatography for both organic—for example, high performance liquid chromatography—and inorganic residues—for example, ion chromatography.

Since the quality of food and the sustainability of the processes are paramount in the 21st century, the chemical analysis provides reliable information to the agroindustry and to the consumer to achieve both goals.

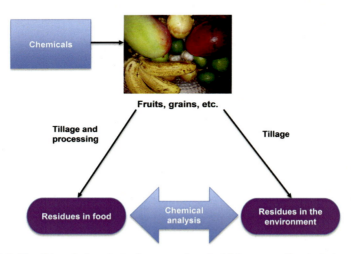

Figure 1.3 The fate of chemicals (i.e., agrochemicals) in an agrifood chain with the use of chemical analysis to determine them. Tillage and processing steps can be considered as the main sources of chemical residues to be analyzed. *Credit: author.*

1.1 Definition of agricultural residues and chemical residues

The definition of *agricultural residues* is large and heterogeneous due to the fact that it derives from *biomass*, a larger kind of analytical matrix (to be consider in Chapter 4: Fundamentals of Analytical Chemistry).

Biomass can be defined as "material produced by the growth of microorganisms, plants, or animals" (International Union of Pure and Applied Chemistry, 2020). Then, agricultural residues can be understood as "the material derived from biomass without any use or interesting in a certain moment of the steps of harvest, processing, and consumption."

Regarding to the *chemical residues*, they can be understood as a chemical species with a contaminant activity for a certain processing or use. Furthermore, a chemical residue can present a toxicological or ecotoxicological effect to living beings (plants, animals, and microorganisms). For instance, the processing of agricultural products, as grains, can generate residues in the liquid, solid or gaseous state:
- Liquids: pesticides (e.g., herbicides, insecticides and fungicides) in aqueous solution in water, soil, and parts of plants;
- Solids: plastics and microplastics from packages;
- Gaseous: carbon dioxide (CO_2) from combustion of fossil fuels.

These residues can achieve the environment by means agricultural practices in agricultural systems (Fig. 1.2). Animals and humans are exposed to these residues by means of food and feed consumption.

Furthermore, it is expected that modern society becomes increasingly aware of the quality of food and committed to the conservation of the environment, which tends to increase the demand for the analysis of these classes of chemical residues.

1.2 Main organic chemical residues

Considering organic molecules—some of them commonly known as POPs (persistent organic pollutants) and another as emerging pollutants (EPs) —the main source comes from pesticides application during the tillage step. Moreover, if we are considering livestock products, veterinary drugs (e.g., antibiotics and anti-inflammatories) are the main residues.

Table 1.1 describes examples of organic molecules that can be observed in agricultural and livestock products as chemical residues. It is expected that these molecules can generate residues in the final product and in the environment.

Table 1.1 Sources and fates of some classes of organic residues in the food chain and in the environment.

Class of residue	Source(s)	Main fate on food chain	Main fate on the environment	Example of scientific reference
Pesticides: herbicide (glyphosate)	Tillage	Final consumer	Soil Surface water Groundwater Outdoor(air)	Mamy, Barriuso, and Gabrielle (2016)
Pesticides: insecticide (chlorpyrifos)	Tillage	Final consumer	Soil Surface water Groundwater Outdoor(air)	Mackay, Giesy, and Solomon (2014)
Pesticides: fungicides (azoxystrobin, boscalid, and pyraclostrobin)	Tillage	Final consumer	Soil Surface water Groundwater Outdoor(air)	Smalling et al. (2013)
Veterinary drugs: antibiotic (oxytetracycline)	Livestock	Final consumer	Soil Surface water Groundwater	Vaz, Lopes, and Martin-Neto (2015)
Veterinary drugs: anti-inflammatory (flunixin)	Livestock	Final consumer	Soil Surface water Groundwater	Popova and Morra (2020)
Plastics: microplastic (plastic pieces with size less than 5 mm)	Packing	Final consumer	Soil Surface water Groundwater	Wu et al. (2019)
Emerging pollutants	Several	Final consumer	Soil Surface water Groundwater Outdoor(air)	Vaz Jr (2018)

Regarding specifically to organic chemical residues found in food from tillage and processing there is a theme of heated discussions because it deals with public health conditions and legislations, and with an expected risk of human exposure. A case example is depicted in Table 1.2 for an

Table 1.2 Results of analytical detection of organic pesticide residues—as active ingredient—in rice (cereal); grape and guava (fruits with edible peels); orange, pineapple, mango (fruits with nonedible peels); lettuce (leafy vegetable); tomato, chayote and peppers (nonleafy vegetables); garlic, beet, sweet potatoes and carrot (roots, tubers and bulbs); and coffee. 122 detected pesticides from 270 searched

Position	Active ingredient	Total of detections
1st	Imidacloprid	713
2nd	Tebuconazole	570
3rd	Carbendazim	526
4th	Pyraclostrobine	522
5th	Dithiocarbamates	464
6th	Difenoconazole	415
7th	Acephate	318
8th	Procymidone	297
9th	Cypermethrine	258
10th	Azoxystrobin	251

Reproduced with permission from: Brazilian National Health Surveillance Agency (2019) Pesticide residue analysis program (in Portuguese). http://portal.anvisa.gov.br/documents/111215/0/Apresentacao + - + PARA_dez_2019.pdf/6321e60d-5910-4a61-9e3d-79a2602ebafa, accessed August 2020.

analytical investigation of 270 pesticide residues in cereal, fruits, vegetables, roots, tubers and bulbs, and coffee.

Fig. 1.4 shown some chemical structures of compounds cited in Tables 1.1−1.2.

1.3 Main inorganic chemical residues

Unlike organic residues, inorganic chemical residues are a limited class observed in food but well-observed in the environment mainly due to the large use of inorganic fertilizers in tillage, specifically for the soil preparation.

However, it is possible to find some inorganic residues in food detected by analytical methods (Fera Science Limited, 2020):

- Inorganic bromide derived from the use of methyl bromide as a postharvest fumigant for stored products or as a soil sterilant.
- Some toxic metals such as arsenic, lead, cadmium, and mercury, occur naturally in certain foods, because they are environmental pollutants in air, water and soil and enter the food supply when plants take them up as they grow and therefore are heavily regulated.
- Iodine is a trace element that is essential for health which is naturally present in some foods, and may be added to others to increase levels

8 Analysis of Chemical Residues in Agriculture

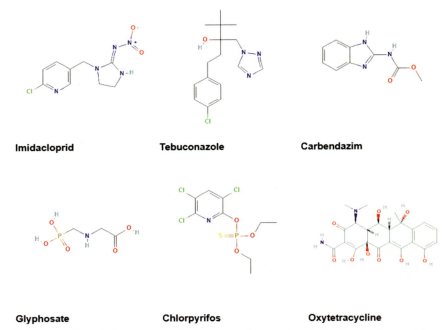

Figure 1.4 Chemical structures of some organic agrochemicals observed as residues on food (imidacloprid, tebuconazole, carbendazim) and on the environment (glyphosate, chlorpyrifos, oxytetracycline). *Credit: author.*

present. Levels of iodine consumption are directly linked to a range of health issues, most notably thyroid conditions. However, there is a safe threshold amount and if iodine intakes are excessive, excess iodine can cause damage to the thyroid.
- Inorganic arsenic, which is known to be far more toxic than its organic form is naturally present within groundwater across the globe which can have a profound effect on the food safety of crops such as rice.

Regarding the effect on the environment of some inorganic chemicals applied in tillage, we can highlight the case of inorganic phosphorus. According to Ali et al. (2019), despite phosphorus being quite abundant in many soils, it is largely unavailable for plant uptake because it forms insoluble complexes with cations under acid and alkaline conditions. As a result, a large amount of inorganic phosphorus fertilizers has been applied for agricultural production systems. Nevertheless, an excessive phosphorus fertilization contributes to greenhouse gases and has a direct negative impact on surface waters.

1.4 Degradation processes[1]

From their chemical structures and their environmental fate, the reactions suffered by organic and inorganic agrochemicals are *hydrolysis, photolysis, oxidation*, and *reduction*. Hydrolysis is defined as the solvolysis by the water, or a reaction with a solvent (water) involving the rupture of one or more bonds in the reacting solute (e.g., pesticide molecule) (International Union of Pure and Applied Chemistry, 2020). Photolysis is the cleavage of one or more covalent bonds in a molecular entity—once again, a pesticide molecule—resulting from absorption of light, or a photochemical process in which such cleavage is an essential part (International Union of Pure and Applied Chemistry, 2020). Oxidation involves the loss of one or more electrons to an oxidative species whereas reduction involves the gain of one or more electrons from a reductive species—these are common processes for metallic species because their electroactive property.

Hydrolysis (Scheme 1.1) occurs when the agrochemical molecule is in aqueous and vapor phase—with or without the presence of an acid—and photolysis (Scheme 1.2) when the molecule is exposed to the electromagnetic radiation ($h\nu$), as UV and visible radiation. These are the main reactions associated with the degradation of pollutants in the environmental matrixes (Balmer, Goss, & Schwarzenbach, 2000; Don, Qiang, Lian, & Qu, 2017).

$$(CH_3)_3N + H_2O \rightarrow (CH_3)_3NH^+ + OH^-$$

$$HNO_2 + h\nu \rightarrow OH + NO$$

Regarding to the kinetic of these reactions, Carlos et al. (2012) observed a pseudo-first order for instance EPs (e.g., clofibric acid) photolysis.

Degradation of a certain molecule under environmental condition—for example, a POP in soil or air—is a frequent process that involves, mainly,

(CH₃)₃N + H₂O → (CH3)₃NH⁺ + OH⁻

Scheme 1.1 The hydrolysis reaction of the trimethylamine, a basic nitrogen compound.

HNO₂ + *hν* → ˙OH + NO

Scheme 1.2 The photolysis reaction of the nitrous acid in the indoor atmosphere.

---

[1] Part of this item was based on: Vaz Jr S, Degradation processes of EPs. In: Vaz Jr S, Analytical chemistry applied to emerging pollutants, p. 103–105, Cham, Springer Nature, 2018

biodegradation and *photodegradation*. In the first case, the International Union of Pure and Applied Chemistry (2020) defines it as the breakdown of a substance catalyzed by enzymes in vitro or in vivo. This may be characterized for purposes of hazard assessment as:

1. Primary: alteration of the chemical structure of a substance resulting in loss of a specific property of that substance.
2. Environmentally acceptable: biodegradation to such an extent as to remove undesirable properties of the compound. This often corresponds to primary biodegradation but it depends on the circumstances under which the products are discharged into the environment.
3. Ultimate: complete breakdown of a compound to either fully oxidized or reduced simple molecules (such as carbon dioxide/methane, nitrate/ammonium and water). It should be noted that the products of biodegradation can be more harmful than the substance degraded.

Photodegradation is defined by the International Union of Pure and Applied Chemistry (2020) as the photochemical transformation of a molecule into lower molecular weight fragments, usually in an oxidation process. This term is widely used in the destruction (oxidation) of pollutants by UV-based processes.

Biodegradation is most common for soil and water and related analytical matrices (e.g., sewage, wastewater, etc.) while photodegradation is most common for the air or atmosphere.

The photodegradation of EPs (e.g., nonsteroidal anti-inflammatory drugs) in surface water at $\lambda > 290$ nm depends on season, pH value, and presence of humic acids and nitrate ion (Koumaki et al., 2015)—humic substances and/or nitrate/nitrite ions can serve as photosensitizers. Apart from photodegradation, the fate of such chemicals in the aquatic environment may be influenced by hydrolysis reactions in an unclear mode (Koumaki et al., 2015).

Regarding to the biodegradation processes, they can involve (Manahan, 2000):
- nitro group reduction
- alkene reduction
- disulphide reduction
- sulfoxide reduction
- ketone reduction
- aldehyde reduction

1.5 Metabolites

A metabolite is any intermediate or product resulting from metabolism, that is the entire physical and chemical processes involved in the maintenance and reproduction of life in which nutrients are broken down to generate energy and to give simpler molecules (catabolism) which by themselves may be used to form more complex molecules (anabolism) (International Union of Pure and Applied Chemistry, 2020).

From this definition we can understand that a certain chemical residue—organic or inorganic—when consumed by a human living can suffer reactions involved in its metabolism generating its metabolite or a molecule derived from them. For instance, the herbicide glyphosate (N-phosphonomethylglycine) can generate two metabolites: aminomethylphosphonic acid and methylphosphonic acid (Fig. 1.5).

Taking into account an organochlorine pesticide as dichlorodiphenyltrichloroethane, its metabolites are dichlorodiphenyldichloroethylene and dichlorodiphenyldichloroethane (Fig. 1.6).

For analytical studies related to regulatory purposes it is recommended the determination not only the agrochemical but their metabolites also because the metabolite could be more dangerous than the precursor.

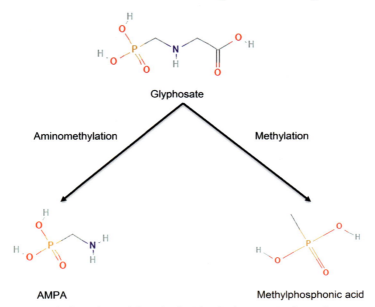

Figure 1.5 Metabolites derived from herbicide glyphosate. *Credit: author.*

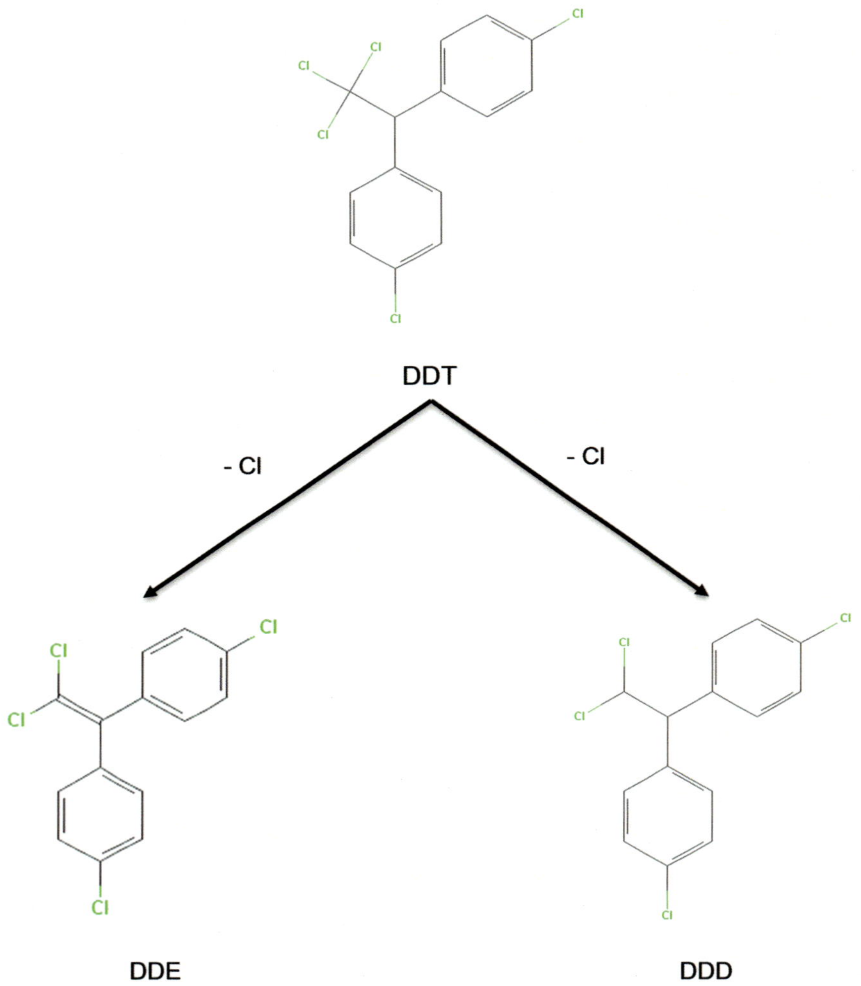

Figure 1.6 Metabolites derived from insecticide DDT. *DDT, Dichlorodiphenyltrichloroethane. Credit: author.*

1.6 International control policies for residues in crops and food

According to the World Health Organization (2020), there are more than 1000 pesticides used around the world to ensure food is not damaged or destroyed by pests. Each pesticide and other agrochemicals have different properties and toxicological effects—which will be discussed in depth in Chapter 3, Toxicology in Agriculture.

International policy efforts are paramount to ensure the quality of agricultural systems—considering the whole chain, from crops to food—their products and their sustainability, highlighting:
- Stockholm Convention[2]—an international treaty that aims to eliminate or restrict the production and use of POPs.
- Joint FAO/WHO Meeting on Pesticide Residues[3]—an independent, international expert scientific group that conduct risk assessments for pesticide residues in food. Moreover, aspects of environmental pollution are considered also.
- Codex Alimentarius Commission[4]—an intergovernmental standards-setting body to establish maximum residue limits (MRLs) for pesticides in food.
- International Code of Conduct on Pesticide Management[5]—The World Health Organization and FAO joint venture to guide government regulators, the private sector, civil society, and other stakeholders on best practices in managing pesticides throughout their lifecycle—from production to disposal.

1.7 Food safety and standardization

Food security and the establishment of quality standards are also paramount to ensure the quality for food. From this scenario, the Codex Alimentarius Commission (CAC) can be understood as the main global effort dedicated to construct and maintain a safety world agricultural trade.

As an example of Codex standard, the document CXS 193-1995 deals with the general standard for contaminants and toxins in food and feed, listing the maximum levels and associated sampling plans of contaminants and natural toxicants in food and feed which are recommended by the CAC to be applied to commodities moving in international trade (Codex Alimentarius, 2019). Some examples of contaminants are:
- copper
- pesticide residues
- residues of veterinary drugs

[2] http://www.pops.int/
[3] https://www.who.int/foodsafety/areas_work/chemical-risks/jmpr/en/
[4] http://www.fao.org/fao-who-codexalimentarius/committees/cac/about/en/
[5] http://www.fao.org/agriculture/crops/thematic-sitemap/theme/pests/code/en/

- microbial toxins, such as botulinum toxin and staphylococcus enterotoxin, and microorganisms
- residues of processing or additives

Special attention is dedicated to the analytical methods as a toll to monitoring and determining the presence or absence of these contaminants.

Table 1.3 describes some examples of values for MRL in food. MRL is commonly used to control the presence of these agrochemicals in

Table 1.3 Maximum residue limit (MRL) for some food products according the Codex Alimentarius (2020).

Pesticide	Functional class	MRL (mg kg^{-1})
2,4-D	Herbicide	0.01 (eggs, milks)
		0.05 (sugarcane, maize)
		2 (wheat)
Acephate	Insecticide	0.01 (eggs)
		0.02 (milks)
		0.05 (meat)
Azoxystrobin	Fungicide	0.01 (eggs, milks)
		0.5 (rice)
		15 (citrus fruits)
Carbendazim	Fungicide	0.05 (wheat)
		0.1 (coffee beans)
		0.5 (tomato, soya bean)
Cypermethrine	Insecticide	0.01 (eggs, onion, root and tuber vegetables)
		2 (barley, meat)
		15 (tea)
Difenoconazole	Fungicide	0.1 (banana, onion, soya bean)
		0.2 (carrot, meat, papaya)
		8 (rice)
Dithiocarbamates	Fungicide	0.5 (garlic, lettuce, melons)
		1 (barley, carrot, peppers)
		5 (grapes, papaya, strawberry)
Imidacloprid	Insecticide	0.02 (poultry meat)
		0.5 (apple, broccoli, root and tuber vegetables)
Pyraclostrobine	Fungicide	0.01 (asparagus, cacao)
		0.02 (banana, tuberous and corm vegetables)
		0.3 (pineapple, tomato)
Tebuconazole	Fungicide	0.15 (wheat, soya bean, peanut)
		1 (apple, peppers)
		6 (grapes)

products for final consumption. In order to establish a complementary way for possible environmental impacts. Table 1.4 describes guidelines values for chemicals from agricultural activities that are of health significance in drinking water as the environmental matrix of relevance for analytical measurements.

Both Tables 1.3 and 1.4 corroborate to the necessity of analytical processes to determining and understanding the effects of the presence of

Table 1.4 Guideline values for chemicals from agricultural activities that are of health significance in drinking water.

Chemical	Guideline value ($\mu g\ L^{-1}$)	Remarks
Nonpesticides		
Nitrate (as NO_3^-)	50,000	Based on short-term effects, but protective for long-term effects
Nitrite (as NO_2^-)	3000	Based on short-term effects, but protective for long-term effects
Pesticides used in agriculture		
Alachlor	20[a]	
Aldicarb	10	Applies to aldicarb sulfoxide and aldicarb sulfone
Aldrin and dieldrin	0.03	For combined aldrin plus dieldrin
Atrazine and its chloro-striazine metabolites	100	
Carbofuran	7	
Chlordane	0.2	
Chlorotoluron	30	
Chlorpyrifos	30	
Cyanazine	0.6	
2,4-D[b]	30	Applies to free acid
2,4-DB[c]	90	
1,2-Dibromo-3-chloropropane	1[a]	
1,2-Dibromoethane	0.4[a](P)	
1,2-Dichloropropane	40 (P)	
1,3-Dichloropropene	20[a]	
Dichlorprop	100	
Dimethoate	6	
Endrin	0.6	
Fenoprop	9	

(Continued)

Table 1.4 (Continued)

Chemical	Guideline value ($\mu g\, L^{-1}$)	Remarks
Hydroxyatrazine	200	Atrazine metabolite
Isoproturon	9	
Lindane	2	
Mecoprop	10	
Methoxychlor	20	
Metolachlor	10	
Molinate	6	
Pendimethalin	20	
Simazine	2	
2,4,5-T[d]	9	
Terbuthylazine	7	
Trifluralin	20	

P, provisional guideline value because of uncertainties in the health database.
[a] For substances that are considered to be carcinogenic, the guideline value is the concentration in drinking water associated with an upper-bound excess lifetime cancer risk of $10-5$ (one additional cancer per 100,000 of the population ingesting drinking water containing the substance at the guideline value for 70 years). Concentrations associated with estimated upper-bound excess lifetime cancer risks of $10-4$ and $10-6$ can be calculated by multiplying and dividing, respectively, the guideline value by 10.
[b] 2,4-Dichlorophenoxyacetic acid.
[c] 2,4-Dichlorophenoxybutyric acid.
[d] 2,4,5-Trichlorophenoxyacetic acid.
Reproduced with permission from World Health Organization (2017). Guidelines for drinking-water quality, 4th edition. WHO, Geneva.

chemical residues in food and in the environment because the modern agricultural systems are very intensive in the use of agrochemicals. Moreover, livestock also contribute to this supply of contaminants and pollutants based on, mainly, the large use of veterinary drugs.

1.8 New trends in agrochemistry

Agrochemicals can be understood as a large class of agricultural inputs that comprise:
- fertilizers
- plant growth regulators
- phytosanitary products, pesticides or correctives (e.g., herbicides, insecticides, fungicides, acaricides, bactericides, nematicides, disinfectants, antibiotics, defoliants, algaecides, repellents)

Despite their relevance to crop protection, they offer risks to the public health and to the environment. Then more sustainable agrochemicals are desirable and are subject of research & development and publications (Vaz Jr, 2019). In order to achieve this requirement, new trends of agrochemistry are related to:
- aspects of food security, that is, the food availability to the populations;
- aspects of food safety, that is, good processing practices to prevent food contamination and foodborne illness;
- aspects of sustainability, that is, environmental, societal and economic impacts;
- aspects of analytical and environmental chemistry in order to monitoring and control contaminants and pollutants[6]. Additionally, the treatment of them;
- Biotechnology, such as, CRISPR gene-editing technology (Shew, Nalley, Snell, Nayga, & Dixon, 2018);
- Nanotechnology, such as, controlled release systems of active ingredients and nanofertilizers (Singh et al., 2020);
- Green chemistry, such as, more safer agrochemicals (Perlatti, Forim, & Zuin, 2014).

From these aspects, a reduction is expected on the negative impacts on public health and on the environment, highlighting analytical and environmental chemistry subject.

1.9 Conclusion

Agriculture is a large economic activity that comprises the production of raw material for food, feed, chemicals, pharmaceuticals, among other products. From this point of view, agricultural practices and food chains need chemical substances (agrochemicals) to production and processing. However, these chemicals produce organic and inorganic residues that need to be analyzed in order to ensure food quality and to avoid environmental pollution.

---

[6] Note: contaminants are minor impurity present in a certain material. Pollutants are undesirable chemical species occurring in the environment as a result of human activities, causing adverse impacts.

References

Ali, W., Nadeem, M., Ashiq, W., Zaeem, M., Shah, S., Gilani, M., ... Cheema, M. (2019). The effects of organic and inorganic phosphorus amendments on the biochemical attributes and active microbial population of agriculture podzols following silage corn cultivation in boreal climate. *Scientific Reports, 9*, 17297.

Balmer, M. E., Goss, K.-U., & Schwarzenbach, R. P. (2000). Photolytic transformation of organic pollutants on soil surfaces-an experimental approach. *Environmental Science & Technology, 34*, 1240–1245.

Brazilian National Health Surveillance Agency (2019). Pesticide residue analysis program (in Portuguese). http://portal.anvisa.gov.br/documents/111215/0/Apresentacao + - + PARA_dez_2019.pdf/6321e60d-5910-4a61-9e3d-79a2602ebafa, Accessed August 2020.

Carlos, L., Mártire, D. O., Gonzalez, M. C., Gomis, J., Bernabeu, A., Amat, A. M., & Arques, A. (2012). Photochemical fate of a mixture of emerging pollutants in the presence of humic substances. *Water Research, 46*, 4732–4740.

Codex Alimentarius (2019). General standard for contaminants and toxins in food and feed. CXS 193-1995. http://www.fao.org/fao-who-codexalimentarius/sh-proxy/tr/?lnk = 1&url = https%253A%252F%252Fworkspace.fao.org%252Fsites%252Fcodex%252FStandards%252FCXS%2B193-1995%252FCXS_193e.pdf, Accessed August 2020.

Codex Alimentarius (2020). Pesticide index. http://www.fao.org/fao-who-codexalimentarius/codex-texts/dbs/pestres/pesticides/en/, Accessed August 2020.

Don, H., Qiang, Z., Lian, J., & Qu, J. (2017). Degradation of nitro-based pharmaceuticals by UV photolysis: Kinetics and simultaneous reduction on halonitromethanes formation potential. *Water Research, 119*, 83–90.

Fera Science Limited (2020). Analytical chemistry services. https://www.fera.co.uk/, Accessed August 2020.

Food and Agriculture Organization of the United Nations (2018). Statistical pocketbook 2018. http://www.fao.org/3/CA1796EN/ca1796en.pdf, Accessed August 2020.

International Union of Pure and Applied Chemistry (2020). Gold book. https://goldbook.iupac.org/terms/view/B00660, Accessed August 2020.

Koumaki, E., Mamais, D., Noutsopoulos, C., Nika, M.-C., Bletsou, A. A., Thomaidis, N. S., ... Stratogianni, G. (2015). Degradation of emerging contaminants from water under natural sunlight: The effect of season, pH, humic acids and nitrate and identification of photodegradation by-products. *Chemosphere, 138*, 675–681.

Mackay, D., Giesy, J. P., & Solomon, K. R. (2014). Fate in the environment and long-range atmospheric transport of the organophosphorus insecticide, chlorpyrifos and its oxon. In J. Giesy, & K. Solomon (Eds.), *Ecological risk assessment for chlorpyrifos in terrestrial and aquatic systems in the United States. Reviews of environmental contamination and toxicology (continuation of residue reviews)* (vol 231). Cham: Springer. Available from https://doi.org/10.1007/978-3-319-03865-0_3.

Mamy, L., Barriuso, E., & Gabrielle, B. (2016). Glyphosate fate in soils when arriving in plant residues. *Chemosphere, 154*, 425–433.

Manahan, S. E. (2000). *Environmental chemistry*. Boca Racton: CRC Press.

Perlatti, B., Forim, M. R., & Zuin, V. G. (2014). Green chemistry, sustainable agriculture and processing systems: A Brazilian overview. *Chemical and Biological Technologies in Agriculture, 1*(5). Available from https://doi.org/10.1186/s40538-014-0005-1.

Popova, I. E., & Morra, M. J. (2020). Fate of the nonsteroidal, anti-inflammatory veterinary drug flunixin in agricultural soils and dairy manure. *Science and Pollution Research International, 16*, 19746–19753.

Shew, A. M., Nalley, L. L., Snell, H. A., Nayga, E. M., Jr, & Dixon, B. L. (2018). CRISPR versus GMOs: Public acceptance and valuation. *Global Food Security, 19*, 71–80.

Singh, A., Dhiman, N., Kar, A. K., Singh, M., Gosh, D., & Patnaik, S. (2020). Advances in controlled release pesticide formulations: Prospects to safer integrated pest management and sustainable agriculture. *Journal of Hazardous Materials*, *385*, 121525.

Smalling, K. L., Kuivila, K. M., Orlando, J. L., Phillips, B. M., Anderson, B. S., Siegler, K., ... Hamilton, M. (2013). Environmental fate of fungicides and other current-use pesticides in a central California estuary. *Marine Pollution Bulletin*, *73*, 144−153.

Vaz, S., Jr (2018). *Analytical chemistry applied to emerging pollutants*. Chan: Springer Nature.

Vaz, S., Jr (2019). *Sustainable agrochemistry − A compendium of technologies*. Chan: Springer Nature.

Vaz, S., Jr, Lopes, W. T., & Martin-Neto, L. (2015). Study of molecular interactions between humic acid from Brazilian soil and the antibiotic oxytetracycline. *Environmental Technology & Innovation*, *4*, 260−267.

World Health Organization. (2017). *Guidelines for drinking-water quality* (4th edition). Geneva: WHO.

World Health Organization (2020). Pesticide residues in food. https://www.who.int/news-room/fact-sheets/detail/pesticide-residues-in-food, Accessed August 2020.

Wu, P., Huang, J., Zheng, Y., Yang, Y., Zhang, Y., He, F., ... Ao, B. (2019). Environmental occurrences, fate, and impacts of microplastics. *Ecotoxicology and Environmental Safety*, *184*, 109612.

CHAPTER 2
Agricultural matrices

Agricultural and related matrices comprise a large diversity of organic and inorganic materials for chemical analysis to determine presence or absence of chemical residues. Despite their heterogeneous compositions, these matrices can be studied by means of several classes of analytical techniques such as chromatography, spectroscopy, and electrochemistry.

Environmental chemistry, as the branch of the chemical sciences that studies the processes involved with the dynamic of the chemical species (e.g., molecules, and ions), can provide information to understand the fate and dynamics of these residues in the environment. On the other hand, food chemistry can provide information to understand also the fate and dynamics of these residues in the food supply and in the human beings. Both scientific branches can contribute to achieve reliable analytical results—that is, the correct interpretation of the generated analytical result. In this context, it is very important to define a sample—a part of a certain material to be analyzed—as: the matrix (the medium to be analyzed) plus the analyte (the chemical species of interest for the analysis) (Fig. 2.1). Then it is easier to establish the correct analytical approach, that is, the best methodology to be applied. Furthermore, sampling is the first

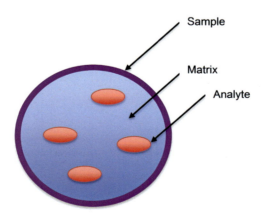

Figure 2.1 The illustration of a sample constitution. *Credit: author.*

experimental step to obtain the correct portion to be analyzed. These concepts will be treated in depth in Chapter 4, Fundamentals of Analytical Chemistry.

2.1 Environmental matrices—soil and water

Environmental matrices are directly impacted by the use and application of agrochemicals into agricultural and livestock systems being these matrices are a receptacle for chemical residues in the environment. Fig. 2.2 depicts this concept based on the entrance of agrochemicals in the production chains.

As discussed in Chapter 1, Introduction to Organic and Inorganic Residues in Agriculture, agricultural systems are highly intensive in the

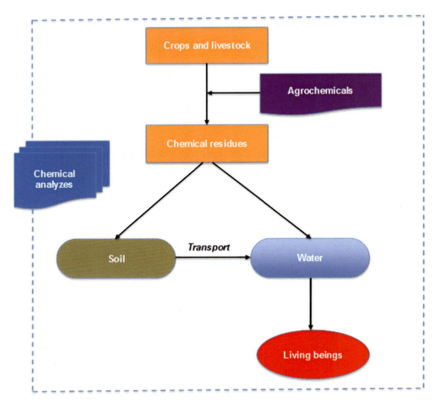

Figure 2.2 The entrance of agrochemicals in the environmental matrices with the application of chemical analyzes for monitoring and control. *Credit: author.*

use of agrochemicals; within seconds chemical residues can be generated in food and in the environment which can reach the soil and water and, in the last step, reach living beings. To monitor and control the presence of these chemical residues, chemical analyzes are paramount.

For the purposes of this book, two environmental matrices will be exploited: soil and water, as follows.

2.1.1 Soil

Soil is the most complex environmental matrix due to the chemical constitution of its organic and inorganic components, and their physical states—soil is formed by chemical substances in the solid, liquid, and gaseous states. Then, soils have a natural tendency to interact with different chemical species, among them some pollutants (e.g., POPs, persistent organic pollutants).

The distribution of organic matter (OM) and the mineral fraction in layers in the soil is as follows:
- Horizon O (surface or topsoil): OM in decomposition (0.3 m of depth).
- Horizon A: OM accumulated mixed with the mineral fraction (0.6 m of depth).
- Horizon B: clay accumulation, Fe^{3+}, Al^{3+} and low OM content (approximately 1 m of depth).
- Horizon C: materials from rock mother.

Thus it is expected that the higher the concentration of OM, especially of humic acids present in it, the greater the retention capacity of metallic cations in soils, especially in the horizon O, which leads to a reduction in the transport of metallic species in the soil, as the humic substances act as strong complexing agents due to the presence of binder sites formed by carboxylic and phenolic groups (Clapp, Hayes, Senesi, Bloom, & Jardine, 2001). Therefore a higher concentration, for example, of bivalent metal cations in the samples of horizons O and A of the soil is expected, considering the effect of the presence of silicate compounds in the metal retention, where a greater value of cation exchange capacity (CEC) of the soil denotes a higher availability of binding sites for the metal after the exit of cations or protons associated with these silicates, due to the negative surface charge of the latter. It is worth mentioning that these physicochemical properties are closely related to the soil functionality.

24 Analysis of Chemical Residues in Agriculture

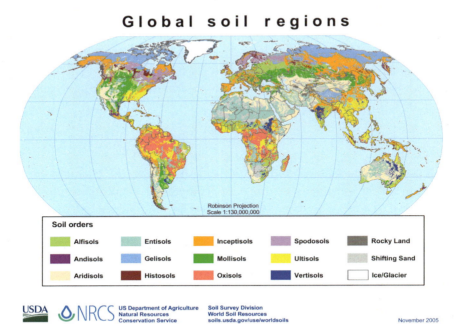

Figure 2.3 The global soil distribution. *Reproduced with permission from United States Department of Agriculture. (2020). Global soil regions map. <https://www.nrcs.usda.gov/wps/portal/nrcs/detail/soils/use/worldsoils/?cid = nrcs142p2_054013> Accessed 8.20. Credit: author.*

Fig. 2.3 depicts a map with the global soil distribution according to the soil classification (alfisols, andisols, aridisols, entosols, histosols, inceptsols, mollisols, oxisols, spodosols, ultisols, vertisols, rocky land, shifting sand, and ice/glacier) (United States Department of Agriculture, 2020). More information about this global distribution can be reached on FAO[1] and USDA[2] websites.

Regarding to the presence of pesticides residues in soils, Fig. 2.4 depicts a map of distribution of these chemical species in European soils (Silva et al., 2019). It is possible to observe the presence of residues in a large portion of Europe due to agricultural activities.

Without a massive application of chemical analyzes—for example, by means chromatographic techniques—it would not possible determine these polluting agents.

---

[1] http://www.fao.org/soils-portal/soil-survey/soil-maps-and-databases/faounesco-soil-map-of-the-world/en/
[2] https://www.nrcs.usda.gov/wps/portal/nrcs/main/soils/use/worldsoils/

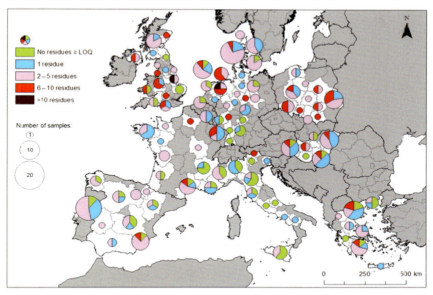

Figure 2.4 A map of distribution of pesticide residues in several European countries: 76 residues of pesticides were analyzed in 317 EU agricultural topsoils (horizon O); glyphosate and its metabolite AMPA, DDTs (DDT and its metabolites) and the broad-spectrum fungicides boscalid, epoxiconazole, and tebuconazole were the compounds most frequently found in soil samples and the compounds found at the highest concentrations. LOQ, limit of quantification. *Reproduced with permission from Elsevier, Silva, V., Mol, H. G. J., Zomer, P., Tienstra, M., Ritsema, C. J., & Geissen, V. (2019). Pesticides residues in European agricultural soils—a hidden reality unfolded.* Science of the Total Environment, 653, 1532–1545.

2.1.2 Water

When we talk about water as an environmental matrix, we must consider it in the plural, since we are dealing with two distinct types, but which strongly correlate: *surface water* and *groundwater*. The surface water is that found in rivers, lakes, seas, and oceans; while groundwater is that found in the aquifers. Drinking-water and wastewater are classifications related to surface and groundwater according to their use.

Regarding to its physicochemical properties, Snoeyink and Jenkins (1996) observed that groundwater, which has a higher concentration of CO_2 gas, is in greater contact with rocks and soil, which leads to a longer dissolution time. The carbonic acid (H_2CO_3) produced by the solubilization of CO_2 when in contact with these materials leads to the solubilization of the minerals, releasing their constituent ions.

A large amount of suspended material can be found mainly in surface waters. Clay, sand, and OM are examples of particles in suspension. There is also a large amount of microorganisms present in water, highlighting bacteria such as coliforms and cyanobacteria, which often compromise the quality of water, especially surface water.

Important analytical parameters for the monitoring of water quality are:

- Electrical conductivity (EC): it provides information on the distribution of ionic species in the medium, with the conductivity being directly proportional to the concentration of these species.
- Dissolved oxygen (DO): O_2 gas has a low solubility in water, with a reduction of its concentration indicating its consumption by the chemical oxygen demand (COD) for the formation of oxidized species, as well as consumption by biochemical demand (BOD) due to the activity of the metabolism of present microorganisms—groundwater have DO values much smaller than surface waters.
- pH scale: its value indicate the concentration of H^+ in the medium—a pH value around six is the most common in drinking-waters; however, there are variations due to the presence of organic or inorganic species.
- Redox potential (E_h): indicates the oxidizing or reducing characteristic of the medium, having a direct correlation with the pH values.
- Presence of organic compounds: determination of petroleum derivatives, agrochemicals and organochlorine compounds produced by treatment processes, which are the main xenobiotics—that is, chemicals found but not produced in organisms or in the environment (Soucek, 2011)—observed in waters.
- Presence of toxic metals: cadmium, mercury, chromium, etc.—are also, in most cases, xenobiotic species.

Unfortunately, water bodies—especially surface water —— are the main route of exposition to pollution due to sanitary problems, such as the absence of wastewater and sewage treatment.

Other water-related matrices are:

- Sediments: naturally occurring material, as rocks, sand and silt, in contact with water bodies, as rivers.
- Sewage or municipal wastewater.
- Sludge: a residual semi-solid material from industrial, water or wastewater treatment processes.

- Wastewater: result of a domestic, industrial, commercial or agricultural activities, with negative impacts to the human health and environment.

We can observe that water is frequently polluted by agroindustrial activities as uncontrolled use of pesticides, discharge of effluents from food processing, among others, demanding a wastewater treatment and reuse in agriculture (Food and Agriculture Organization of the United Nations, 2020), according to microbiological and chemical analyses (World Health Organization, 2006).

Table 2.1 show several categories of water pollutants from agricultural activities, as crops, livestock, and aquaculture (Food and Agriculture Organization of the United Nations and the International Water Management Institute, 2017).

2.2 Food and feed

Self (2005) observed that the natural origins of human foods are biologically diverse, ranging widely in texture and composition—from nutmeg to oysters. The extremely complex endogenous composition of food is made even more complex in the modern environment where so many extrinsic, additional items—additives such as antioxidants, contaminants from agriculture such as herbicides, and industrial adulterants such as hydrocarbons from petroleum—may also be present. Regarding food, feed can be seen also as a highly heterogeneous matrix. Of course the subjects here are those contaminants—or residues directly generated—from agricultural practices and systems.

Table 2.2 features the chemical composition of some food in order to present their heterogeneous composition for a chemical analysis (FooDB, 2020).

It is expected that crops with more intensive pesticides use, for example, soybean and corn, can present higher residue concentrations than those less intensive crops, such as lettuce and garlic.

Regarding livestock feed, we can consider also soybean and corn crops, plus sorghum (*Sorghum bicolor*)—commonly used for the manufacture of animal feed—as the main sources of pesticide residues due to the high demand of these agrochemicals for the tillage and harvest steps. They are relevant agricultural matrices from this point of view based on a previous tendency of residue presence in the grain. For instance, we have two formulations of dairy cattle feed:

Table 2.1 Categories of major water pollutants in agriculture and the relative contributions of the three main agricultural production systems.

Pollutant category	Indicators/examples	Relative contribution by Crops	Livestock	Aquaculture
Nutrients	Primarily nitrogen and phosphorus present in chemical and organic fertilizers as well as animal excreta and normally found in water as nitrate, ammonia or phosphate	***	***	*
Pesticides	Herbicides, insecticides, fungicides and bactericides, including organophosphates, carbamates, pyrethroids, organochlorine pesticides and others (many, such as DDT, are banned in most countries but are still being used illegally and persistently)	***	—	—
Salts	For example, ions of sodium, chloride, potassium, magnesium, sulphate, calcium and bicarbonate. Measured in water, either directly as total dissolved solids or indirectly as electric conductivity	***	*	*
Sediment	Measured in water as total suspended solids or nephelometric turbidity units—especially from pond drainage during harvesting	***	***	*
Organic matter	Chemical or biochemical oxygen demanding substances (e.g., organic materials such as plant matter and livestock excreta), which use up dissolved oxygen in water when they degrade	*	***	**
Pathogens	Bacteria and pathogen indicators. For example, *Escherichia coli*, total coliforms, fecal coliforms and enterococci	*	***	*
Metals	For example, selenium, lead, copper, mercury, arsenic and manganese	*	*	*
Emerging pollutants	For example, drug residues, hormones and feed additives	—	***	**

Reproduced with permission from Food and Agriculture Organization of the United Nations and the International Water Management Institute. (2017). *Water pollution from agriculture: A global review.* <http://www.fao.org/3/a-i7754e.pdf> Accessed 8.20.

Table 2.2 Chemical composition of common food; some of them are also used for feed.

Macronutrients	Content range (mg/100 g) Lettuce (Lactuca sativa)	Pineapple (Ananas comosus)	Garlic (Allium sativum)	Oat (Avena sativa)	Papaya (Carica papaya)
Ash	0.36000–800.00000	0.20000–400.00000	1.510–3300.000	0.000–2300.000	0.15000–500.00000
Carbohydrate	1200.000–4000.000	10700.000–15500.000	31000.000–72700.000	57500.000–69100.000	9900.000–12000.000
Fat	100.000–400.000	100.000–400.000	500.000–800.000	6400.000–6500.000	300.000–300.000
Fatty acids	0.000–200.000	0.000–200.000	0.000–400.000	0.000–2700.000	100.000–100.000
Fiber (dietary)	0.000–2100.000	0.000–2200.000	0.000–9900.000	0.000–11500.000	0.000–2100.000
Proteins	1000.000–1400.000	300.000–500.000	6400.000–16800.000	13000.000–13900.000	700.000–700.000
Water	0.60000–95700.00000	4.535–85700.000	58.580–58600.000	6.550–84000.000	4.220–89165.000

Macronutrients	Content range (mg/100 g) Lemon (Citrus limon)	Sweet orange (Citrus sinensis)	Arabica coffee (Coffea arabica)	Soybean (Glycine max)	Barley (Hordeum vulgare L.)	Corn (Zea mays)
Ash	0.19000–600.00000	0.20000–2500.00000	8.880–10400.000	0.000–5100.000	0.27000–1500.00000	0.000–2200.000
Carbohydrate	4300.000–16000.000	9500.000–49700.000	0.000–85940.000	0.000–34000.000	64400.000–76900.000	0.000–88800.000
Fat	0.000–1100.000	100.000–600.000	0.000–15400.000	200.000–100000.000	2400.000–3000.000	600.000–100000.000
Fatty acids	0.000–500.000	0.000–200.000	0.000–29100.000	0.000–60500.000	200.000–1400.000	0.000–56200.000
Fiber (dietary)	0.000–10600.000	0.000–2000.000	0.000–19800.000	0.000–16600.000	0.000–9500.000	0.000–8100.000
Proteins	400.000–1500.000	500.000–900.000	0.000–19500.000	0.000–88320.000	9200.000–9200.000	0.000–10830.000
Water	88.980–92600.000	1.200–87000.000	5500.000–99400.000	0.000–90400.000	10.090–88300.000	0.000–86670.000

The objective of this table is to present a general description of the chemical constitution—or composition—of these matrices. For detailed information access the FooDB website: https://www.foodb.ca/

Reproduced with permission from FooDB. (2020). *Listing foods*. https://foodb.ca/foods > Accessed 8.20.

- Concentrated with 18% m/m CP (crude protein): 68.8% m/m ground corn; 27.2% m/m soybean meal; and 4% m/m of core (mineral/vitamin);
- Concentrated with 24% m/m CP: 53.1% m/m ground corn; 42.9% m/m soybean meal; and 4% m/m of core (mineral/vitamin).

Moreover, Brambilla and De Filippis (2005) proposed the addition of other components in order to improve feed and food safety, and established some analytical challenges related to their use and control (Table 2.3).

The analytical challenges described in Table 2.3 can be faced by means of the best analytical approach. In this case, chromatographic techniques such as LC-MS and GC-MS (further seen in Chapter 5: Main Analytical Techniques) are the best option.

2.3 Meat and other animal products

According Lawrie and Ledward (2006), the chemical meat composition (m/m) consists of approximately:
- 75% of water
- 19% of protein
- 2.5% of fat
- 1.2% of carbohydrates
- 1.65% of other nitrogen compounds

It also contains a great amount of several inorganic species of Ca, P, Na, K, Cl, Mg, and trace elements such as Fe, Cu, Zn, and many others. Then it is a highly heterogeneous analytical matrix for itself.

Table 2.4 describes the composition of chicken, cattle, and swine meats.

Residues are expected to be found in meat from veterinary drugs such as antibiotics (e.g., macrolides, sulphonamides, nitroimidazoles, tetracyclines) (Fig. 2.5). Furthermore, some pesticides residues can be detected in livestock and poultry excrement (Wang et al., 2020), which can be seen as an indicator of their presence or passing through the animal organism.

As a practical example that generates the demand for chemical analysis, European commission established that "*food-producing animals may be treated with veterinary medicines to prevent or cure disease. These substances may leave residues in the food from treated animals. Food may also contain residues of pesticides and contaminants to which animals have been exposed to. In all cases, the levels of residues in food should not harm the consumer*" (European Commission, 2020). From this statement, we can infer that it can be extended to:

Table 2.3 Summary of the analytical challenges foreseen as consequences of the changes in animal feeds composition. PUFA, polyunsaturated fatty acids.

Feed component	Possible consequences	Analytical challenges
Colostrum as immunological product	Intake of hormones naturally present in pregnant animals in males and/or pre-puberal ones	Characterization of the hormonal profiles, with respect to the presence of the polar second phase metabolites
Algae and other sea derived materials as PUFA renewable sources	Intake of bio-toxins, and "sea drugs"	Study of their metabolisms and possible bioaccumulation in terrestrial farmed animals and related food commodities
Long chain polyunsaturated fatty acids as nutraceuticals	Increased levels of endogenous cannabinoids, such as Anandamide	Monitoring of the natural levels in farmed animals and their products
Engineerised probiotics, to improve feed conversion	Fermentation of vegetal alimentary substrate, to produce steroids hormones and their precursors	Characterization of the fermentation products under strictly anaerobic condition, with respect also to the formation of esters as hydrophobic second phase metabolites
Herbal extracts as alternative to anti-microbials	Presence of natural pharmacological/biological active substances	Identification and determination of the active principles, along to the study of their metabolic fate. Correlation between their activity and the amount present in the extracts

The objective of this table is to present a general description of the chemical constitution—or composition—of these matrices. For detailed information access the FooDB website: https://www.foodb.ca/.
Reproduced with permission from, Brambilla, G. & De Filippis, S. (2005). Trends in animal feed composition and the possible consequences on residue tests. *Analytica Chimica Acta, 529*, 7—13.

- Cheese
- Eggs
- Honey
- Milk

And their composition can be seen in the Table 2.5.

Table 2.4 Chemical composition of some meats.

Macronutrients	Content range (mg 100 g^{-1}) Chicken	Cattle	Swine
Ash	0.000–3400.000	0.10000–5100.00000	0.000–6000.000
Carbohydrate	0.000–9400.000	0.000–11000.000	0.000–13000.000
Fat	2700.000–17400.000	1600.000–30100.000	1000.000–51000.000
Fatty acids	700.000–7100.000	0.000–94000.000	0.000–22900.000
Fiber (dietary)	0.000–300.000	0.000–1800.000	0.000–400.000
Proteins	15600.000–28300.000	1500.000–36710.000	10000.000–58000.000
Water	1.050–1.050	4000.000–84160.000	1700.000–78600.000

The objective of this table is to present a general description of the chemical constitution—or composition—of these matrices. For detailed information access the FooDB website: https://www.foodb.ca/.
Reproduced with permission from, FooDB. (2020). *Listing foods*. https://foodb.ca/foods> Accessed 8.20.

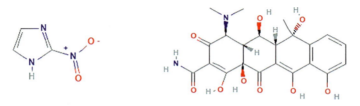

Azithromycin (a macrolide) **Sulfanilamide (a sulphonamide)**

2-Nitroimidazole (a nitroimidazole) **Oxyteracycline (a tetracycline)**

Figure 2.5 Some structures of antibiotic molecules applied as veterinary drugs.

2.4 Beverages and fruit juices

Beverages and fruit juices are, basically, aqueous matrices with several compounds diluted, as can be seen in Table 2.6.

Wang et al. (2019) observed the transfer of pesticide residue during tea brewing and proposed that it could be extended to other beverage crops (e.g., grape and other fruits and barley).

2.5 Agroindustrial waste

Agroindustrial waste is generated from crops and livestock processing. It is a high-content OM material—a source of biomass—and chemically a very heterogeneous matrix which depends on:
- Raw material composition (or constitution)
- Processing steps
- Final product composition

It is not a common subject of analysis from regulatory agencies, and it is very important understand that any pesticide residue from tillage or

Table 2.5 Chemical composition of some animal products, except meat.

Macronutrients	Content range (mg 100 g^{-1})			
	Cheese	Egg	Honey	Milk (dairy cattle or cow)
Ash	0.40000–8000.00000	0.60000–10600.00000	52.000–52.000	0.79000–800.00000
Carbohydrate	200.000–16000.000	300.000–11500.000	82120.000–82120.000	4500.000–4500.000
Fat	400.000–44100.000	0.000–55800.000	—	4100.000–4100.000
Fatty acids	0.000–28500.000	0.000–21000.000	—	200.000–2600.000
Fiber (dietary)	—	—	—	—
Proteins	3500.000–38500.000	10500.000–81100.000	—	3600.000–3600.000
Water	0.70000–87600.00000	1.000–87800.000	17.100–17100.000	87000.000–87000.000

The objective of this table is to present a general description of the chemical constitution—or composition—of these matrices. For detailed information access the FooDB website: https://www.foodb.ca/.

Reproduced with permission from, FooDB. (2020). *Listing foods*. https://foodb.ca/foods > Accessed 8.20.

Table 2.6 Chemical composition of some beverages and juices.

Macronutrients	Content range (mg 100 g^{-1}) Grape wine	Beer	Fruit juice (e.g., orange juice)
Ash	0.13000–300.00000	0.06000–200.00000	0.01000–200.00000
Carbohydrate	200.000–5900.000	2700.000–8600.000	8300.000–42000.000
Ethanol	3300.000–78000.000	0.04000–4.80000	—
Proteins	100.000–300.000	200.000–400.000	200.000–200.000
Water	83400.000–96000.000	91.960–94900.000	89690.000–95200.000

The objective of this table is to present a general description of the chemical constitution—or composition—of these matrices. For detailed information access the FooDB website: https://www.foodb.ca/.
Reproduced with permission from, FooDB. (2020). *Listing foods*. https://foodb.ca/foods> Accessed 8.20.

harvest—for crops—and any veterinary drug—for animal products—can be present here.

Examples of waste chemical composition include pig slurry and waste of beer fermentation. The first comes from an animal (pig) source and the second from a crop (barley) source in order to demonstrate the diversity of agroindustrial waste.

2.5.1 Pig slurry

According Hunce, Clemente, and Bernal (2020), the chemical composition of a pig slurry in dry basis is:***
- Moisture (%wt./wt.): 62.7 ± 7.04
- pH: 7.49 ± 0.25
- EC (dS m^{-1}): 6.10 ± 0.82
- OM (%wt./wt.): 32.0 ± 5.14
- Water soluble carbon (g kg^{-1}): 8.0 ± 0.27
- Total organic carbon (g kg^{-1}): 381 ± 19.2
- Total nitrogen (g kg^{-1}): 22.1 ± 1.90
- C/N: 17.3 ± 1.79
- Soluble carbohydrates (g kg^{-1}): 2.7 ± 0.0

2.5.2 Waste of beer fermentation

According Khan, Hyon, and Park (2007), the chemical composition of a waste of beer fermentation broth is:
- Carbon (g L^{-1}): 68.114
- Nitrogen (g L^{-1}): 17.178
- Hydrogen (g L^{-1}): 116.141
- Total carbohydrate (g L^{-1}): 29.442
- Total protein (g L^{-1}): 0.671
- Ethanol (%v/v): 4.58

2.6 Conclusion

Agricultural matrices are very heterogeneous regarding their origin and chemical composition. In order to facilitate their understanding, they can be divided into environmental, food and feed, meat and other animal products, beverages and fruit juices, and agroindustrial residues. The presence of chemical residues in these matrices is subject of control by means regulatory agencies in order to attend technical criteria for environmental and food safety.

To apply the best analytical approaches is paramount for the knowledge of the composition and physicochemical properties of these matrices. Then it is possible to obtain reliable analytical results.

References

Brambilla, G., & De Filippis, S. (2005). Trends in animal feed composition and the possible consequences on residue tests. *Analytica Chimica Acta, 529*, 7−13.

Clapp, C. E., Hayes, M. H. B., Senesi, N., Bloom, P. R., & Jardine, P. M. (Eds.), (2001). *Humic substances and chemical contaminants*. Madison: Soil Society of America.

European Commission. (2020). Residues of veterinary medicinal products. <https://ec.europa.eu/food/safety/chemical_safety/vet_med_residues_en> Accessed 9.20.

Food and Agriculture Organization of the United Nations. (2020). Wastewater treatment and reuse in agriculture. <http://www.fao.org/land-water/water/water-management/wastewater/en/> Accessed 8.20.

Food and Agriculture Organization of the United Nations and the International Water Management Institute. (2017). Water pollution from agriculture: A global review. <http://www.fao.org/3/a-i7754e.pdf> Accessed 8.20.

FooDB. (2020). Listing foods. https://foodb.ca/foods> Accessed 8.20.

Hunce, S. Y., Clemente, R., & Bernal, M. P. (2020). Selection of Mediterranean plants biomass for the composting of pig slurry solids based on the heat production during aerobic degradation. *Waste Management, 104*, 1−8.

Khan, T., Hyon, S. H., & Park, J. K. (2007). Production of glucuronan oligosaccharides using the waste of beer fermentation broth as a basal medium. *Enzyme and Microbial Technology, 42*, 89−92.

Lawrie, R. A., & Ledward, D. (2006). *Lawrie's meat science* (7th ed.). Cambridge: Woodhead Publishing.

Self, R. (2005). Methodology and proximate analysis. In R. Self (Ed.), *Extraction of organic analytes from foods: A manual of methods*. Cambridge: Royal Society of Chemistry.

Silva V., Mol H. G. J., Zomer P., Tienstra M., Ritsema C. J., & Geissen V. (2019). Pesticides residues in European agricultural soils—a hidden reality unfolded. *Science of the Total Environment 653*, (pp. 1532−1545).

Snoeyink, V. L., & Jenkins, D. (1996). *Química del água [Water chemistry]*. México City: Limusa.

Soucek, P. (2011). Xenobiotics. In M. Schwab (Ed.), *Encyclopedia of cancer*. Berlin, Heidelberg: Springer. Available from https://doi.org/10.1007/978-3-642-16483-5_6276>.

United States Department of Agriculture. (2020). Global soil regions map. <https://www.nrcs.usda.gov/wps/portal/nrcs/detail/soils/use/worldsoils/?cid=nrcs142p2_054013> Accessed 8.20.

Wang, J., Xu, J., Ji, X., Wu, H., Yang, H., Zhang, H., ... Qian, M. (2020). Determination of veterinary drug/pesticide residues in livestock and poultry excrement using selective accelerated solvent extraction and magnetic material purification combined with ultra-high-performance liquid chromatography−tandem mass spectrometry. *Journal of Chromatography A, 1617*, 460808.

Wang, X., Zhou, L., Zhang, X., Luo, F., & Chen, Z. (2019). Transfer of pesticide residue during tea brewing: Understanding the effects of pesticide's physico-chemical parameters on its transfer behavior. *Food Research International, 121*, 776–784. Available from https://doi.org/10.1016/j.foodres.2018.12.060.

World Health Organization. (2006). *WHO guidelines for the safe use of wastewater, excreta and greywater—v. 2. Wastewater use in agriculture*. Geneva: World Health Organization.

CHAPTER 3
Toxicology in agriculture

Ecotoxicology and toxicology are both interdisciplinary scientific branches that study the effects of chemical substances on the environment (the first) and human beings (the second) in order to define limits for the exposition to those chemicals—for example, pesticides and veterinary drugs. It involves chemistry, biochemistry, biology, pharmacy, and medicine being chemical analyzes tools to access the necessary data to formulate and to confirm the required scientific hypothesis behind the establish limits or concentrations for monitoring and control.

Chemical analyzes—especially those conducted by means the use of chromatographic, spectroscopic and electrochemical techniques—are paramount to obtain data to estimate ecotoxicological and toxicological parameters (e.g., LD_{50} and LC_{50}) related to the presence of chemical residues from agriculture in several analytical matrixes (e.g., food, soil, water, among others). Fig. 3.1 illustrates a generic flowchart for ecotoxicological and toxicological purposes considering the risk evaluation for the exposure to a certain chemical.

In a simple way, a risk evaluation methodology can comprise sampling, analytical parameters and maximum values, analysis itself, data quality and control. A chemical residue identification comprises the quantitative and/or qualitative data for the target-compound. The exposure evaluation or assessment describes the duration and frequency of exposure to a certain chemical. The dose-response evaluation describes the association between dose and a biological effect. The risk characterization comprises the hazard identification. Finally, the risk management comprises decision-making processes based on the determined previous information. All steps follow official regulatory actions and technical guidelines to guarantee the effectiveness of their proceedings.

Terms presented in the Fig. 3.1 are explored in detail in the next item.

3.1 Definition of main terms in ecotoxicology and toxicology
3.1.1 Ecotoxicology

The terms presented here were adapted from IUPAC Recommendations (Nordberg, Templeton, Andersen, & Duffus, 2009) in order to offer a general vision of the relevance of chemical analyzes in ecotoxicology.

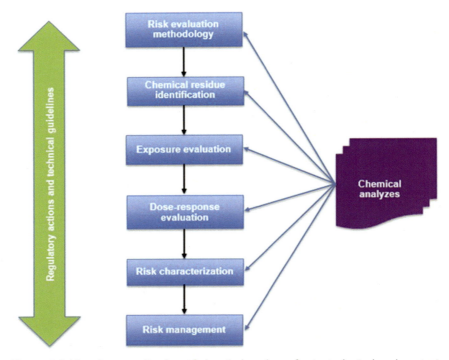

Figure 3.1 The direct application of chemical analyzes for toxicological and ecotoxicological purposes. *Credit: Sílvio Vaz Jr.*

Abiotic—Not associated with living organisms.

Abiotic transformation—Process in which a substance in the environment is modified by nonbiological mechanisms.

Absorbate—Substance that enters and is retained inside a solid or semisolid matrix (absorbent).

Absorbent—Solid or semisolid matrix that is able to accommodate and retain an absorbate. See also sorbate, sorbent.

Absorption (in general)—(1) Process of one material (absorbate) being retained by another (absorbent). Note: The process may be the physical solution of a gas, liquid, or solid in a liquid, attachment of molecules of a gas, vapor, liquid, or dissolved substance to a solid surface by physical forces, etc. See also adsorption. (2) Transfer of some or all of the energy of radiation to matter which it traverses. Note: Absorption of light at bands of characteristic wavelengths is used as an analytical method in spectrophotometry to identify the chemical nature of molecules, atoms, or ions and to measure the concentrations of these species.

Acute—(1) Of short duration, in relation to exposure or effect; (2) the effect usually shows a rapid onset.

Adsorbate—Molecular species of gas, dissolved substance, or liquid that adheres to or is adsorbed in an extremely thin surface layer of a solid substance.

Adsorbent—Condensed phase at the surface of which adsorption may occur.

Adsorption—Increase in the concentration of a substance at the interface of a condensed layer and a liquid or a gaseous layer owing to the operation of surface forces.

Adsorption factor—Ratio of the amount of substance adsorbed at the interface of a condensed layer and a liquid or gaseous phase to the total amount of the substance available for adsorption.

Ambient monitoring—Continuous or repeated measurement of agents in the environment to evaluate ambient exposure and health risk by comparison with appropriate reference values based on knowledge of the probable relationship between exposure and resultant adverse health effects.

Analysis plan (in ecological risk assessment)—Scheme that defines the exact format and design of the assessment, explicitly states the data needed, and describes the methods and design for analyzing these data.

Antagonism—Combined effect of two or more factors that is less than that expected from simple summation of toxicities of the individual compounds.

Anthropogenic—(1) Caused by or influenced by human activities. (2) Describing a conversion factor used to calculate a dose or concentration affecting a human that has been derived from data obtained with another species, for example, the rat.

Baseline toxicity—General, nonspecific, reversible mode of toxic action that can be produced in most living organisms by the presence of sufficient amounts of many organic chemicals. Note 1: Effects result from the general disruption of cellular activity. The mechanism producing disruption is unknown, with the main theories being binding to proteins in cell membranes and "swelling" of the lipid portion of cell membranes resulting from the presence of organic chemicals. Note 2: Hydrophobicity dominates the expression of baseline toxicity.

Bioaccessibility—Potential for a substance to come in contact with a living organism and perhaps interact with it, with the possibility of absorption into the organism. In ecotoxicology, bioaccessibility is often measured by assessment of the fraction of a substance released from a matrix (usually soil or sediment) into an aqueous medium under defined laboratory conditions. Such measurements must be interpreted with care as laboratory conditions rarely equate to those in nature. In ecotoxicology, bioaccessibility is often measured by assessment of the fraction of a substance released from a matrix (usually soil or sediment) into an aqueous medium under defined laboratory conditions. Such measurements must be interpreted with care as laboratory conditions rarely equate to those in nature.

Bioaccumulation—Progressive increase in the amount of a substance in an organism or part of an organism that occurs because the rate of intake from all contributing sources and by all possible routes exceeds the organism's ability to eliminate the substance from its body.

Bioavailability, F (in toxico- or pharmacokinetics)—Ratio of the systemic exposure from extravascular (ev) exposure to that following intravenous (iv) exposure as described by the Eq. (3.1):

$$F = \frac{A\text{ev } D\text{iv}}{B\text{iv } D\text{ev}} \qquad (3.1)$$

where F (fraction of dose absorbed) is a measure of the bioavailability, A and B are the areas under the concentration–time curve following ev and iv administration, respectively, and $D\text{ev}$ and $D\text{iv}$ are the administered ev and iv doses.

Biodegradation—Breakdown of a substance catalyzed by enzymes in vitro or in vivo. In ecotoxicology, it is almost entirely due to microbial or fungal activity. Note 1: Biodegradation may be classified for purposes of hazard assessment into three categories: (1) Primary. Alteration of the chemical structure of a substance resulting in loss of a specific property of that substance. (2) Environmentally acceptable. Biodegradation to such an extent as to remove undesirable properties of the compound. This often corresponds to primary biodegradation but depends on the circumstances under which the products are discharged into the environment. (3) Ultimate. Complete breakdown of a compound to either fully oxidized or reduced simple molecules (such as carbon dioxide, methane, nitrate, ammonium, and (or) water). Note 2: The products of biodegradation can be more harmful than the substance that was degraded.

Biomarker—Quantifiable behavioral, physiological, histological, biochemical, or genetic property that is used to measure response to an environmental change.

Biosensor—Device that uses specific biochemical reactions mediated by isolated enzymes, immunosystems, tissues, organelles, or whole cells to detect chemical compounds usually by electrical, thermal, or optical signals.

Biotransformation (bioconversion)—Chemical conversion of a substance that is mediated by living organisms or enzyme preparations derived therefrom.

Chemisorption—Sorption which results from chemical bond formation (strong interaction) between the sorbent and the sorbate in a monolayer on a surface or internal to an absorbent.

Contaminant—(1) Minor impurity present in a substance. (2). Extraneous material inadvertently added to a sample prior to or during chemical analysis. (3) In some contexts, as in relation to gas cleaning equipment, used as a synonym for "pollutant", especially on a small scale. (4). Unintended component in food that may pose a hazard to the consumer. (5) Any undesirable solid, liquid, or gaseous matter occurring, as a result of human activities, in a solid, liquid, or gaseous environmental medium, even without adverse effects being observed.

Degradation (breakdown, decomposition)—Process by which a substance is broken down to simpler structures through biological or abiotic mechanisms.

Dose–response relationship—Association between dose and the incidence of a defined biological effect in an exposed population usually expressed as percentage.

Ecology—Branch of biology that studies the interactions between living organisms and all factors (including other organisms) in their environment. Such interactions encompass environmental factors that determine the distributions of living organisms.

Ecosystem—Grouping of organisms (microorganisms, plants, animals) interacting together, with and through their physical and chemical environments, to form a functional entity within a defined environment.

Ecotoxicology—Study of the toxic effects of chemical and physical agents on all living organisms, especially on populations and communities within defined ecosystems; it includes transfer pathways of these agents and their interactions with the environment.

Emission (discharge, effluent, release)—Release of a substance from a source, including discharges to the wider environment.

Endocrine disrupter—Exogenous chemical that alters function(s) of the endocrine system and consequently causes adverse health effects in an intact organism, its progeny, or (sub) populations.

Endogenous—Produced within or caused by factors within an organism.

Environmental assessment (EA)—Short, preliminary assessment of potential environmental harm used to determine if a full environmental impact statement (EIS) is required.

Environmental availability—Portion of the total amount of a substance present in the environment that is involved in a particular process and is subject to physicochemical and biological modifying influences. See also environmental bioavailability.

Environmental bioavailability—Ratio of uptake clearance to the rate at which an organism encounters a given contaminant in an environmental medium (e.g., soil, sediment, water, food) being processed by the organism. Note: This is a measure of an organism's extraction efficiency, via respiratory, dietary, and surface absorption processes, from the environmentally available (bioaccessible) portion of a material.

Environmental fate—Destiny of a chemical or biological pollutant after release into the natural environment.

Environmental impact assessment (EIA)—Appraisal of the possible environmental consequences of a past, ongoing, or planned action, resulting in the production of an EIS or "finding of no significant impact" (FONSI).

Environmental monitoring—Continuous or repeated measurement of agents in the environment to evaluate environmental exposure and possible damage by comparison with appropriate reference values based on knowledge of the probable relationship between ambient exposure and resultant adverse effects.

Environmental risk assessment—Estimate of the probability that harm will result from a defined exposure to a substance in an environmental medium. The estimate is valid only for a given species and set of conditions.

Eutrophic—Describing an environment having a high concentration of nutrients. Note: The term is usually used to describe nutrient-rich bodies of water or soil having a high or excessive rate of biological production.

Eutrophication—Process producing a high concentration of nutrient salts and a high or excessive rate of biological production in water. Note: Usually involves depletion of the oxygen content caused by decay of organic matter resulting from high primary production as a result of enhanced input of nutrients.

Exposure assessment—Process of measuring or estimating concentration (or intensity), duration, and frequency of exposures to an agent present in the environment or, if estimating hypothetical exposures, that might arise from the release of a substance, or radionuclide, into the environment.

Exposure pathway—Route by which an individual is exposed to a contaminant or pollutant, including the source and point of contact.

Fate—Disposition of a material in various environmental compartments (e.g., soil or sediment, water, air, biota) as a result of transport, partitioning, transformation, and degradation.

Freundlich adsorption isotherm—Empirical equation that describes the adsorption of a contaminant to soil. The Eq. (3.2) is:

$$\frac{x}{m} = (K_f C_e)\exp\left(\frac{1}{n}\right) \tag{3.2}$$

where x/m is the mass ratio of concentrations of adsorbed substance at equilibrium divided by the concentration of the contaminant in soil, C_e is the contaminant (or pollutant) concentration in the aqueous phase at equilibrium, K_f is the equilibrium constant (the Freundlich adsorption constant), and $1/n$ is a contaminant specific exponent.

Good laboratory practice (GLP) (principles)—Fundamental rules incorporated in Organization for Economic Cooperation and Development (OECD) guidelines and national regulations concerned with the process of effective organization and the conditions under which laboratory studies are properly planned, performed, monitored, recorded, and reported.

Greenhouse effect—The net warming of the Earth resulting from increasing atmospheric concentrations of carbon dioxide (CO_2), water vapor, and other greenhouse gases. Note: Gases and vapors such as CO_2 and water vapor are relatively transparent to light but absorb long-wave, infrared radiation radiating back from the Earth's surface. The net balance for sunlight influx, infrared radiation absorption by greenhouse gases, and infrared efflux from the Earth's surface determines the steady-state temperature of the Earth.

Greenhouse gases—Atmospheric gases that are relatively transparent to sunlight entering the atmosphere but absorb infrared radiation generated at the Earth's surface. They include water vapor, carbon dioxide (CO_2), methane, dinitrogen oxide (nitrous oxide), chlorofluorocarbons (CFCs), chloroethane (methylchloroform), carbon tetrachloride, and the fire retardant, halon. Ozone in the troposphere can also act as a greenhouse gas.

Hazard—Set of inherent properties of a substance, mixture of substances, or a process involving substances that, under production, usage, or disposal conditions, make it capable of causing adverse effects to organisms or the environment, depending on the degree of exposure; in other words, it is a source of danger. See also risk.

Hazard assessment—Determination of factors controlling the likely effects of a hazard such as the dose−effect and dose−response relationships, variations in target susceptibility, bioaccumulation potential, persistence, and mechanism of toxicity.

Hazard evaluation—Identification and assessment of the potential adverse effects that could result from manufacture, use, and disposal of a material in a specified quantity and manner.

Inhibitory concentration (IC)—Concentration of a substance that causes a defined inhibition of a given system. Note: IC_{50} is the median concentration that causes 50% inhibition in a nonlethal biological measurement of the test organisms, such as reproduction or growth.

Lethal concentration (LC)—Concentration of a substance in an environmental medium that causes death following a certain period of exposure. Note: LC_{50} is the median concentration that causes death in 50% of the test population.

Lethal dose (LD)—Amount of a substance or physical agent (e.g., radiation) that causes death when taken into the body. Note: LD_{50} is the median dose that causes death in 50% of the test population.

Lethal time (LT)—Time taken for a defined percentage, usually 50%, of a test population to die. Note: The median lethal time (MLT) for 50% of the test population is referred to as the MLT_{50}.

Lethality—Ability to cause death.

Limit test—Acute toxicity test in which, if no ill effects occur at a preselected maximum dose, no further testing at greater exposure levels is required.

Lowest effective dose (LED)—Lowest dose of a chemical inducing a specified effect in a specified fraction of exposed individuals.

Lowest-observed-adverse-effect level (LOAEL)—Lowest concentration or amount of a substance (dose), found by experiment or observation, which causes an adverse effect on morphology, functional capacity, growth, development, or life span of a target organism distinguishable from normal (control) organisms of the same species and strain under defined conditions of exposure.

Lowest-observed-effect concentration (LOEC)—Lowest concentration of a material used in an aquatic toxicity test that has a statistically significant adverse effect on the exposed population of test organisms compared with controls. Note: When derived from a life cycle or partial life cycle test, it is numerically the same as the upper limit of the maximum acceptable toxicant concentration (MATC). Also called lowest-observed-adverse-effect level (LOAEL).

Lowest-observed-effect level (LOEL)—Lowest concentration or amount of a substance (dose), found by experiment or observation, that causes any alteration in morphology, functional capacity, growth, development, or life span of target organisms distinguishable from normal (control) organisms of the same species and strain under the same defined conditions of exposure.

Median effective concentration (EC$_{50}$)—Statistically derived median concentration of a substance in an environmental medium expected to produce a certain effect in 50% of test organisms in a given population under a defined set of conditions. Note: EC$_n$ refers to the median concentration that is effective in n% of the test population.

Median effective time (ET$_{50}$)—For sublethal or ambiguously lethal effects, the median time until 50% of the exposed individuals respond.

Median effective concentration (EC$_{50}$)—Statistically derived median concentration of a substance in an environmental medium expected to produce a certain effect in 50% of test organisms in a given population under a defined set of conditions. Note: EC$_n$ refers to the median concentration that is effective in *n*% of the test population.

Median effective dose (ED$_{50}$)—Statistically derived median dose of a chemical or physical agent (radiation) expected to produce a certain effect in 50% of test organisms in a given population or to produce a half-maximal effect in a biological system under a defined set of conditions. Note: ED$_n$ refers to the median dose that is effective in n% of the test population.

Median inhibitory time (IT$_{50}$)—Time required for a toxicant to inhibit a specified process in 50% of a population under test conditions. See also effect time, lethal time.

Median lethal concentration (LC$_{50}$)—Statistically derived median concentration of a substance in an environmental medium expected to kill 50% of organisms in a given population under a defined set of conditions.

Median lethal dose (LD$_{50}$)—Statistically derived median dose of a chemical or physical agent (radiation) expected to kill 50% of organisms in a given population under a defined set of conditions.

Median lethal time (LT$_{50}$)—Statistically derived median time interval during which 50% of a given population may be expected to die following acute administration of a chemical or physical agent (radiation) at a given concentration under a defined set of conditions.

Median teratogenic concentration (TC$_{50}$)—Median concentration resulting in developmental malformations for 50% of the exposed individuals within a predetermined time, for example, 96 hours.

Median time to death (MTTD)—Median time resulting in death for 50% of the exposed organisms.

Median tolerance limit (TLm or TL$_{50}$)—Concentration of a substance in air, water, sediment, or soil at which 50% of the test organisms survive after a specified time of exposure. The TL50 (equivalent to the TLm) is usually expressed as a time-dependent value (e.g., 24- or 96-hours TL$_{50}$; the estimated concentration at which 50% of test organisms survive after 24 or 96 hours of exposure).

Monitoring—Continuous or repeated observation, measurement, and evaluation of health and (or) environmental or technical data for defined purposes, according to prearranged schedules in space and time, using comparable methods for sensing and data collection. Note: Evaluation requires comparison with appropriate reference values based on knowledge of the probable relationship between ambient exposure and adverse effects.

Monitoring test—Test designed to be applied on a routine basis, with some degree of control, to ensure that the quality of an environmental compartment, biological endpoint, or effluent has not exceeded some prescribed criteria range. In a biomonitoring test, organisms are used as "sensors" to detect changes in the quality of water or effluent. A monitoring test implies generation of information, on a continuous or other regular basis.

Monte Carlo simulation—Analysis of a sequence of events using random numbers to generate possible outcomes in an iterative process.

No-observed-adverse-effect level (NOAEL)—Greatest concentration or amount of a substance, found by experiment or observation, which causes no detectable adverse alteration of morphology, functional capacity, growth, development, or life span of the target organism under defined conditions of exposure.

No-observed-effect concentration (NOEC)—Special case of the no-observed-effect level (NOEL), commonly used in aquatic toxicology.

No-observed-effect level (NOEL)—Greatest concentration or amount of a substance, found by experiment or observation, that causes no statistically significant alterations of morphology, functional capacity, growth, development, or life span of target organisms distinguishable from those observed in normal (control) organisms of the same species and strain under the same defined conditions of exposure.

No-response level (NRL)—Maximum dose of a substance at which no specified response is observed in a defined population and under defined conditions of exposure.

Octanol-air partition coefficient, P_{OA}, K_{OA}—Partition coefficient for a compound between octanol and air. Like K_{ow}, it is a measure of lipophilicity.

Octanol-water partition coefficient, P_{ow}, K_{ow}—Ratio of the solubility of a chemical in octanol to its solubility in water at equilibrium. Note: Measure of lipophilicity, used in the assessment of both the uptake and physiological distribution of organic chemicals and prediction of their environmental fate.

Partition coefficient—Concentration of a substance in one phase divided by its concentration in the other phase when the heterogeneous system of two phases is in equilibrium. Note 1: The ratio of concentrations (or, strictly speaking, activities) of the same molecular species in the two phases is constant at constant temperature. Note 2: The octanol-water partition coefficient K_{ow} is often used as a measure of the bioconcentration factor (BCF) for modeling purposes. Note 3: This term is in common usage in toxicology but is not recommended by IUPAC for use in chemistry and should not be used as a synonym for partition constant, partition ratio, or distribution ratio.

Partition ratio, K_D—Ratio of the concentration of a substance in a single definite form, A, in the extract to its concentration in the same form in the other phase at equilibrium, for example, for an aqueous/organic system the Eq. (3.3):

$$K_D(A) = \frac{[A]_{org}}{[A]_{aq}} \qquad (3.3)$$

Pathway—Sequence of enzymatic or other reactions by which one biological material is converted to another.

Persistence—Attribute of a substance that describes the length of time that the substance remains in a particular environment before it is physically removed or chemically or biologically transformed.

Persistent inorganic pollutant (PIP)—Inorganic substance that is stable in the environment, is liable to long-range transport, may bioaccumulate in human and animal tissue, and may have significant impacts on human health and the environment. Note 1: Examples are arsenides, fluorides, cadmium salts, and lead salts. Note 2: Some inorganic chemicals, like crocidolite asbestos, are persistent in almost all circumstances, but others, like metal sulfides, are persistent only in unreactive environments; sulfides can generate hydrogen sulfide in a reducing environment or sulfates and sulfuric acid in oxidizing environments. As with organic substances, persistence is often a function of environmental properties.

Persistent organic pollutant (POP)—Organic chemical that is stable in the environment, is liable to long-range transport, may bioaccumulate in human and animal tissue, and may have significant impacts on human health and the environment. Examples are tetrachlorodibenzodioxin (dioxin), PCBs, and DDT.

Photodegradation—Any breakdown reaction of a chemical compound that is initiated by sunlight (UV light), or more accurately, by the influence of a high-energy photon. This can be either by direct photodegradation, in which the photon photolyzes or ionizes the relevant molecule itself, which then reacts with other species in its vicinity, or by indirect photodegradation, in which the relevant molecule reacts with ions or radicals created by photolysis of other species.

Photo-induced toxicity—Toxicity of a chemical in the presence of light due to the production of toxic photolysis products.

Pollutant—Any undesirable solid, liquid, or gaseous matter occurring, as a result of human activities, in a solid, liquid, or gaseous environmental medium and causing adverse effects. Note 1: "Undesirability", like toxicity, is concentration-dependent, low concentrations of most substances being tolerable or even essential in many cases. Note 2: A primary pollutant is one emitted into the atmosphere, water, sediments, or soil from an identifiable source. Note 3: A secondary pollutant is a pollutant formed by chemical reaction in the atmosphere, water, sediments, or soil. Note 4: Pollutant should be distinguished from contaminant; the latter implies presence above background due to human activities; the former implies that the substance also is causing adverse effects.

Pollution—Introduction of pollutants into a solid, liquid, or gaseous environmental medium; the presence of pollutants in a solid, liquid, or gaseous environmental medium; or any undesirable modification of the composition of a solid, liquid, or gaseous environmental medium.

Population—In ecology, any group of interacting and interbreeding organisms of the same species occupying a given area at the same time.

Predicted no-effect concentration (PNEC)—Concentration that is expected to cause no adverse effect to any naturally occurring population in an environment at risk from exposure to a given substance.

Predictive risk assessment—Risk assessment performed for a proposed future action, such as the use of a new chemical or the release of a new effluent.

Quality standards—Fixed upper limits for exposure to certain chemicals recognized under law by one or more levels of government. Well-known examples include the air, water, and soil quality standards, as well as threshold limit values for air pollutants in the workplace.

Quantitative structure—activity relationship (QSAR)—(1) Quantitative model relating chemical structure of organic compounds to biological activity (including toxicity), derived using regression analysis and containing as parameters physicochemical constants, indicator variables, or theoretically calculated values. Note 1: QSAR is used as a method of predicting toxicity. It is also used to design molecules with a defined biological activity prior to their synthesis for use as drugs, pesticides, etc., and for assessing environmental fate of chemicals. (2) Quantitative model relating chemical structure of compounds to chemical activity in the environment. Note 2: The term is extended by some authors to include chemical reactivity, that is, activity and reactivity are regarded as synonyms. This extension is discouraged.

Quantitative structure—metabolism relationship (QSMR)—Quantitative association between the physicochemical and (or) the structural properties of a substance and its metabolic behavior.

Receptor—Molecular structure in or on a cell that specifically recognizes and binds to a compound and acts as a physiological signal transducer or mediator of an effect.

Reference dose (RfD)—Estimate (with uncertainty spanning perhaps an order of magnitude) of a daily exposure of a defined substance to the human population (including sensitive subgroups) that is likely to be without an appreciable risk of deleterious effects during a lifetime. Note 1: It can be derived from a no-observed-adverse-effect level (NOAEL), lowest-observed adverse-effect level (LOAEL), or benchmark dose, with uncertainty factors generally applied to reflect limitations of the data used. It is generally used in the EPA's noncancer health assessments. Note 2: The RfD is reported in units of mg of substance/kg body weight/day for oral exposures.

Reference material—Material, sufficiently homogeneous and stable regarding one or more properties, used in calibration, in assignment of a value to another material, or in quality assurance.

Reference site—Relatively unpolluted site used for comparison with polluted sites in environmental monitoring studies, often incorrectly referred to as a control site.

Reference toxicant—Chemical used in an aquatic toxicity test as a positive control in contrast to the negative control provided by exposure water without the test chemical. Information collected is used to determine the general health and viability of the test organisms and assess consistency in testing protocol implementation. Note: In this definition, the term "positive control" is used to describe a procedure that is very similar to the actual experimental test and that is known from previous experience to give a positive result.

Reference toxicity test—Test conducted in conjunction with sediment tests to determine possible changes in condition of the test species. Note 1: Deviations from an established normal range indicate a change in the condition of the test organism population. Reference toxicant tests are most often acute lethality tests performed in the absence of sediment. Note 2: Sediment spiked with a toxicant might also be included as a positive control for the sediment toxicity test.

Remediation—Removal of pollution from environmental media such as soil, groundwater, sediment, or surface water for the general protection of human health and the environment.

Resilience—Ability of a community to maintain its structure and function in the face of disturbance, and to reorganize following disturbance-driven change.

Risk—(1) Probability of adverse effects caused under specified circumstances by an agent in an organism, a population, or an ecological system. (2) Probability of a hazard causing an adverse effect. (3) Expected frequency of occurrence of a harmful event arising from such an exposure.

Risk analysis—Process for controlling situations where an organism, system, or subpopulation could be exposed to a hazard. Note 1: The risk analysis process consists of three components: risk assessment, risk management, and risk communication. Note 2: The term is misleading since "analysis" has the fundamental meaning "resolution or breaking up of anything complex into its various simple elements, the opposite process to synthesis; the exact determination of the elements or components of anything complex (with or without their physical separation)."

Risk assessment—Identification and quantification of the risk resulting from a specific use or occurrence of a chemical or physical agent, taking into account possible harmful effects on individuals or populations exposed to the agent in the amount and manner proposed and all the possible routes of exposure. Note 1: Risk assessment is generally considered to involve four steps: hazard identification, hazard characterization, exposure assessment, and risk characterization. Note 2: Quantification ideally requires the establishment of dose−effect and dose−response relationships in likely target individuals and populations.

Risk characterization—Outcome of hazard identification and risk estimation applied to a specific use of a substance or occurrence of an environmental health hazard. Note: Risk characterization requires quantitative data on the exposure of organisms or people at risk in the specific situation. The end product is a quantitative statement about the proportion of organisms or people affected in a target population.

Risk management—Decision-making process involving considerations of political, social, economic, and engineering factors with relevant risk assessments relating to a potential hazard so as to develop, analyze, and compare regulatory options and to select the optimal regulatory response for safety from that hazard. Note: Essentially, risk management is the combination of three steps: risk evaluation; emission and exposure control; and risk monitoring.

Route of exposure—Means by which a toxic agent gains access to an organism by administration through the gastrointestinal tract (ingestion), lungs (inhalation), skin (topical), or by other routes such as intravenous, subcutaneous, intramuscular, or intraperitoneal routes.

Screening—(1) Describing a testing procedure designed to separate people or objects according to a fixed characteristic or property. (2) Carrying out test(s), examination(s), or procedure(s) in order to expose undetected abnormalities, unrecognized (incipient) diseases, or defects: examples are mass X-rays and cervical smears. Note: Pharmacological or toxicological screening consists of a specified set of procedures to which a substance is subjected in order to characterize its pharmacological and toxicological properties and to establish dose—effect and dose—response relationships.

Screening level—Decision limit or cut-off point at which a screening test is regarded as positive.

Screening test (*preliminary test or range-finding test*)—(1). Test conducted to estimate the concentrations to be used for a definitive test. (2). Acute test used early in a testing program to evaluate the potential of a substance to produce a given adverse effect (e.g., mortality).

Sorbate—Noncommittal term used instead of adsorbate or absorbate when the sorption process is undefined.

Sorbent—Noncommittal term used instead of adsorbent or absorbent when the sorption process is undefined.

Sorption—Process whereby a solute becomes physically or chemically associated with a sorbent regardless of the mechanism (absorption, adsorption, chemisorption). Note: Sometimes used instead of adsorption or absorption when it is difficult to discriminate experimentally between these two processes.

Sorption constant—Quantity describing the distribution of a substance between a solvent and a sorbent, typically water and sediment, at equilibrium, according Eq. (3.4):

$$K_d = \frac{[C(\text{sediment})]}{[C(\text{water})]} \tag{3.4}$$

Structure−activity relationship (SAR)—Association between specific aspects of molecular structure and defined biological action. See also quantitative structure−activity relationship.

Surrogate toxicant—Relatively well-studied substance whose properties are assumed to apply to an entire chemically and toxicologically related class; for example, benzo[*a*]pyrene data may be used as toxicologically equivalent to that for all carcinogenic polynuclear aromatic hydrocarbons.

Teratogen—Agent that, following exposure of a mother, may induce uninheritable permanent structural malformations or defects in the offspring.

Teratogenic—Capable of causing uninheritable permanent structural malformations or defects in the offspring of an exposed parent.

Teratogenicity—(1) Potential to cause the production of uninheritable structural malformations or defects in offspring. (2) Production of uninheritable structural malformations or defects in offspring.

Tolerance—(1) Adaptive state characterized by diminished effects of a particular dose of a substance: the process leading to tolerance is called "adaptation". (2) In food toxicology, dose that an individual can tolerate without showing an effect. (3) Ability to experience exposure to potentially harmful amounts of a substance without showing an adverse effect. (4) Ability of an organism to survive in the presence of a toxic substance: increased tolerance may be acquired by adaptation to constant exposure. (5) In immunology, state of specific immunological unresponsiveness.

Top-down ecotoxicological study—Approach to investigating ecotoxicological effects that starts with a determination of the presence and nature of any adverse effects via responses at community and ecosystem levels of organization rather than the suborganismal levels of organization.

Toxic—Able to cause injury to living organisms as a result of physicochemical interaction.

Toxic substance (poison, toxicant, toxic chemical)—Substance causing injury to living organisms as a result of physicochemical interactions. Note 1: All substances are toxic above a certain dose (or exposure). Thus, the term is normally applied only to those substances causing toxicity at relatively low doses. Note 2: Toxicity of any substance varies from organism to organism. Thus, this term should be accompanied by the name of the organism to which it applies, but this is rare. In common use, the term refers to toxicity to humans and related mammals. Note 3: In ecotoxicology, great care should be taken in using this term because of the variation in susceptibility of different species, some of which have adapted to survive, and even benefit from, exposure to substances which are very toxic to many other species.

Toxicity—(1) Capacity to cause injury to a living organism defined with reference to the quantity of substance administered or absorbed, the way in which the substance is administered and distributed in time (single or repeated doses), the type and severity of injury, the time needed to produce the injury, the nature of the organism(s) affected, and other relevant conditions. (2) Adverse effects of a substance on a living organism defined as in 1. (3) Measure of incompatibility of a substance with life. This quantity may be expressed as the reciprocal of the absolute value of median lethal dose ($1/LD_{50}$) or median lethal concentration ($1/LC_{50}$).

Xenobiotic—Compound with a chemical structure foreign to a given organism. Note: Frequently restricted to man-made compounds.

3.1.2 Toxicology

For terms of toxicology, an IUPAC Recommendation (Duffus, Norberg, & Templeton, 2009) also was used as source, which they were adapted when necessary.

Absolute lethal concentration (LC_{100})—Lowest concentration of a substance in an environmental medium which kills 100% of test organisms or species under defined conditions. Note: This value is dependent on the number of organisms used in its assessment.

Absolute lethal dose (LD_{100})—Lowest amount of a substance that kills 100% of test animals under defined conditions. Note: This value is dependent on the number of organisms used in its assessment.

Acceptable daily intake (ADI)—Estimate of the amount of a food additive, expressed on a body weight basis, which can be ingested daily over a lifetime without appreciable health risk. Note 1: For calculation of ADI, a standard body mass of 60 kg is used Note 2: Tolerable daily intake (TDI) is the analogous term used for contaminants.

Acceptable residue level of an antibiotic—Acceptable concentration of a residue that has been established for an antibiotic found in human or animal foods.

Acceptable risk—Probability of suffering disease or injury that is considered to be sufficiently small to be "negligible". Note: Calculated risk of an increase of one case in a million people per year for cancer is usually considered to be negligible.

Accidental exposure—Unintended contact with a substance or change in the physical environment (including, e.g., radiation) resulting from an accident.

Active ingredient (AI)—Component of a mixture responsible for the biological effects of the mixture. Compare inert ingredient.

Active metabolite—Metabolite causing biological and (or) toxicological effects.

Active transport—Movement of a substance across a cell membrane against an electrochemical gradient, in the direction opposite to normal diffusion and requiring the expenditure of energy.

Acute—(1) Of short duration, in relation to exposure or effect; the effect usually shows a rapid onset. Note: In regulatory toxicology, "acute" refers to studies where dosing is either single or limited to one day although the total study duration may extend to two weeks to permit appearance of toxicity in susceptible organ systems. (2) In clinical medicine, sudden and severe, having a rapid onset.

Acute effect—Effect of finite duration occurring rapidly (usually in the first 24 hours or up to 14 d) following a single dose or short exposure to a substance or radiation. Note: Acute effects may occur continuously following continuous dosing or repeatedly following repeated dosing.

Acute exposure—Exposure of short duration.

Acute toxicity—(1) Adverse effects of finite duration occurring within a short time (up to 14 d) after administration of a single dose (or exposure to a given concentration) of a test substance or after multiple doses (exposures), usually within 24 hours of a starting point (which may be exposure to the toxicant, or loss of reserve capacity, or developmental change, etc.). (2) Ability of a substance to cause adverse effects within a short time of dosing or exposure.

Acute toxicity test—Study in which organisms are observed during only a short part of the life span and in which there is often only a single exposure to the test agent at the beginning of the study.

Adsorption—Increase in the concentration of a substance at the interface of a condensed and a liquid or gaseous layer owing to the operation of surface forces.

Adverse effect—Change in biochemistry, physiology, growth, development morphology, behavior, or life span of an organism which results in impairment of functional capacity or impairment of capacity to compensate for additional stress or increase in susceptibility to other environmental influences.

Adverse event—Occurrence that causes an adverse effect. Note: An adverse event in clinical studies is any untoward reaction in a human subject participating in a research project; such an adverse event, which may be a psychological reaction, must be reported to an institutional review board.

Agonist—Substance that binds to cell receptors normally responding to a naturally occurring substance and produces an effect similar to that of the natural substance. Note 1: A partial agonist activates a receptor but does not cause as much of a physiological change as does a full agonist. Note 2: A co-agonist works together with other co-agonists to produce a desired effect.

Ambient monitoring—Continuous or repeated measurement of agents in the environment to evaluate ambient exposure and health risk by comparison with appropriate reference values based on knowledge of the probable relationship between exposure and resultant adverse health effects.

Analytic study (in epidemiology)—Study designed to examine associations, commonly putative or hypothesized causal relationships.

Antagonism—Combined effect of two or more factors that is smaller than the solitary effect of any one of those factors. Note: In bioassays, the term may be used when a specified effect is produced by exposure to either of two factors but not by exposure to both together.

Antagonist—Substance that binds to a cell receptor normally responding to a naturally occurring substance and prevents a response to the natural substance.

Anthropogenic—(1) Caused by or influenced by human activities. (2) Describing a conversion factor used to calculate a dose or concentration affecting a human that has been derived from data obtained with another species (e.g., the rat).

Benchmark concentration (BMC)—Statistically calculated lower 95% confidence limit on the concentration that produces a defined response (called the benchmark response or BMR, usually 5% or 10%) for an adverse effect compared to background, often defined as 0% or 5%.

Benchmark dose (BMD)—Statistically calculated lower 95% confidence limit on the dose that produces a defined response (called the benchmark response or BMR, usually 5% or 10%) of an adverse effect compared to background, often defined as 0% or 5%.

Benchmark guidance value—Biological monitoring guidance value set at the 90th percentile of available biological monitoring results collected from a representative sample of workplaces with good occupational hygiene practices.

Benchmark response—Response, expressed as an excess of background, at which a benchmark dose or benchmark concentration is set.

Bioaccessibility—Potential for a substance to come in contact with a living organism and then interact with it. This may lead to absorption. Note: A substance trapped inside an insoluble particle is not bioaccessible, although substances on the surface of the same particle are accessible and may also be bioavailable. Bioaccessibility, like bioavailability, is a function of both chemical speciation and biological properties. Even surface-bound substances may not be accessible to organisms which require the substances to be in solution.

Bioaccessible—Able to come in contact with a living organism and interact with it.

Bioaccumulation—Progressive increase in the amount of a substance in an organism or part of an organism that occurs because the rate of intake exceeds the organism's ability to remove the substance from the body. Note: Bioaccumulation often correlates with lipophilicity.

Bioaccumulation potential—Ability of living organisms to concentrate a substance obtained either directly from the environment or indirectly through its food.

Bioactivation—Metabolic conversion of a xenobiotic to a more toxic derivative or one which has more of an effect on living organisms.

Bioassay—Procedure for estimating the concentration or biological activity of a substance by measuring its effect on a living system compared to a standard system.

Bioavailability—Extent of absorption of a substance by a living organism compared to a standard system.

Bioavailability, F (in toxico- or pharmacokinetics)—Ratio of the systemic exposure from extravascular (ev) exposure to that following intravenous (iv) exposure as described by the Eq. (3.5):

$$F = \frac{A_{ev} D_{iv}}{B_{iv} D_{ev}} \quad (3.5)$$

where F (fraction of dose absorbed) is a measure of the bioavailability, A and B are the areas under the (plasma) concentration–time curve following extravascular and intravenous administration, respectively, and D_{ev} and D_{iv} are the administered extravascular and intravenous doses.

Bioequivalence—Relationship between two preparations of the same drug in the same dosage form that have a similar bioavailability.

Biokinetics—Science of the movements involved in the distribution of substances.

Biological half-life—For a substance, the time required for the amount of that substance in a biological system to be reduced to one-half of its value by biological processes, when the rate of removal is approximately exponential.

Blood–brain barrier—Physiological interface between brain tissues and circulating blood created by a mechanism that alters the permeability of brain capillaries, so that some substances are prevented from entering brain tissue, while other substances are allowed to enter freely.

Blood–placenta barrier—Physiological interface between maternal and fetal blood circulations that filters out some substances which could harm the fetus while favoring the passage of others such as nutrients: Many fat-soluble substances such as alcohol are not filtered out, and several types of virus can also cross this barrier. Note: The effectiveness of the interface as a barrier varies with species and different forms of placentation.

Carcinogenicity—Process of induction of malignant neoplasms, and thus cancer, by chemical, physical, or biological agents.

Carrier substance—Substance that binds to another substance and transfers it from one site to another.

Chemical safety—Practical certainty that there will be no exposure of organisms to toxic amounts of any substance or group of substances: This implies attaining an acceptably low risk of exposure to potentially toxic substances.

Chronic effect—Consequence that develops slowly and (or) has a long-lasting course; may be applied to an effect which develops rapidly and is long-lasting.

Chronic exposure—Continued exposure or exposures occurring over an extended period of time, or a significant fraction of the test species' or of the group of individuals', or of the population's life-time.

Chronic toxicity—(1) Adverse effects following chronic exposure. (2) Effects that persist over a long period of time whether or not they occur immediately upon exposure or are delayed.

Chronic toxicity test—Study in which organisms are observed during the greater part of the life span and in which exposure to the test agent takes place over the whole observation time or a substantial part thereof.

Clinical toxicology—Scientific study involving research, education, prevention, and treatment of diseases caused by substances such as drugs and toxins. Note: Often refers specifically to the application of toxicological principles to the treatment of human poisoning.

Cohort study—Analytic study in epidemiology in which subsets of a defined population can be identified who are, have been, or in the future may be exposed or not exposed, or exposed in different degrees, to a factor or factors hypothesized to influence the probability of occurrence of a given disease or other outcome. The main feature of the method is observation of a large population for a prolonged period (years), with comparison of incidence rates of the given disease in groups that differ in exposure levels.

Concentration—effect relationship—Association between exposure concentration and the resultant magnitude of the continuously graded change produced, either in an individual or in a population.

Concentration—response curve—Graph of the relation between exposure concentration and the proportion of individuals in a population responding with a defined effect.

Concentration—response relationship—Association between exposure concentration and the incidence of a defined effect in an exposed population.

Contaminant—(1) Minor impurity present in a substance. (2) Extraneous material inadvertently added to a sample prior to or during chemical analysis. (3) In some contexts, as in relation to gas cleaning equipment, used as a synonym for "pollutant", especially on a small scale. (4) Unintended component in food that may pose a hazard to the consumer.

Control group—Selected subjects of study, identified as a rule before a study is done, which comprises humans, animals, or other species who do not have the disease, intervention, procedure, or whatever is being studied, but in all other respects are as nearly identical to the test group as possible.

Critical dose—Dose of a substance at and above which adverse functional changes, reversible or irreversible, occur in a cell or an organ.

Critical effect—For deterministic effects, the first adverse effect that appears when the threshold (critical) concentration or dose is reached in the critical organ: Adverse effects with no defined threshold concentration are regarded as critical.

Cumulative effect—Overall change that occurs after repeated doses of a substance or radiation.

Cumulative median lethal dose—Estimate of the total administered amount of a substance that is associated with the death of half a population of animals when the substance is administered repeatedly in doses which are generally fractions of the median lethal dose.

Cumulative risk—(1) Probability of a common harmful effect associated with concurrent exposure by all relevant pathways and routes of exposure to a group of substances that share a common chemical mechanism of toxicity. (2) Total probability of a harmful effect over time.

Cytochrome P450 (CYP)—Member of a superfamily of heme-containing monooxygenases involved in xenobiotic metabolism, cholesterol biosynthesis, and steroidogenesis, in eukaryotic organisms found mainly in the endoplasmic reticulum and inner mitochondrial membrane of cells. "P450" refers to the observation that a solution of this enzyme exposed to carbon monoxide strongly absorbs light at a wavelength of 450 nm compared with the unexposed solution (a difference spectrum caused by a thiolate in the axial position of the heme opposite to the carbon monoxide ligand).

Cytotoxic—Causing damage to cell structure or function.

Decontamination—Process of rendering harmless (by neutralization, elimination, removal, etc.) a potentially toxic substance in the natural environment, laboratory areas, the workplace, other indoor areas, clothes, food, water, sewage, etc.

Deposition—(1) Process by which a substance arrives at a particular organ or tissue site, for example, the deposition of particles on the ciliated epithelium of the bronchial airways. (2) Process by which a substance sediments out of the atmosphere or water and settles in a certain place.

Dermal irritation—Skin reaction resulting from a single or multiple exposure to a physical or chemical entity at the same site, characterized by the presence of inflammation; it may result in cell death.

Desorption—Decrease in the amount of adsorbed substance; opposite of adsorption.

Detoxication—(1) Process, or processes, of chemical modification that make a toxic molecule less toxic. (2) Treatment of patients suffering from poisoning in such a way as to promote physiological processes which reduce the probability or severity of adverse effects.

Developmental toxicity—Adverse effects on the developing organism (including structural abnormality, altered growth, or functional deficiency or death) resulting from exposure through conception, gestation (including organogenesis), and postnatally up to the time of sexual maturation.

Dose (of a substance)—Total amount of a substance administered to, taken up, or absorbed by an organism, organ, or tissue.

Dose–effect—Relation between dose and the magnitude of a measured biological change.

Dose−effect curve—Graph of the relation between dose and the magnitude of the biological change produced measured in appropriate units.

Dose−effect relationship—Association between dose and the resulting magnitude of a continuously graded change, either in an individual or in a population.

Dose−response curve—Graph of the relation between dose and the proportion of individuals in a population responding with a defined biological effect.

Dose−response relationship—Association between dose and the incidence of a defined biological effect in an exposed population usually expressed as percentage.

Drug (medicine, pharmaceutical)—Any substance that when absorbed into a living organism may modify one or more of its functions. Note: The term is generally accepted for a substance taken for a therapeutic purpose, but is also commonly used for abused substances.

Effective concentration (EC)—Concentration of a substance that causes a defined magnitude of response in a given system. Note: EC_{50} is the median concentration that causes 50% of maximal response.

Effective dose (ED)—Dose of a substance that causes a defined magnitude of response in a given system. Note: ED_{50} is the median dose that causes 50% of maximal response.

Estimated daily intake (EDI)—Prediction of the daily intake of a residue of a potentially harmful agent based on the most realistic estimation of the residue levels in food and the best available food consumption data for a specific population: Residue levels are estimated taking into account known uses of the agent, the range of contaminated commodities, the proportion of a commodity treated, and the quantity of home-grown or imported commodities. Note: The EDI is expressed in mg residue per person.

Estimated exposure concentration (EEC)—Measured or calculated amount or mass concentration of a substance to which an organism is likely to be exposed, considering exposure by all sources and routes.

Estimated exposure dose (EED)—Measured or calculated dose of a substance to which an organism is likely to be exposed, considering exposure by all sources and routes.

Estimated maximum daily intake (EMDI)—Prediction of the maximum daily intake of a residue of a potentially harmful agent based on assumptions of average food consumption per person and maximum residues in the edible portion of a commodity, corrected for the reduction or increase in residues resulting from preparation, cooking, or commercial processing. Note: The EMDI is expressed in mg residue per person.

Exposure—(1) Concentration, amount, or intensity of a particular physical or chemical agent or environmental agent that reaches the target population, organism, organ, tissue, or cell, usually expressed in numerical terms of concentration, duration, and frequency (for chemical agents and microorganisms) or intensity (for physical agents). (2) Process by which a substance becomes available for absorption by the target population, organism, organ, tissue, or cell, by any route. (3) For X- or gamma radiation in air, the sum of the electrical charges of all the ions of one sign produced when all electrons liberated by photons in a suitably small element of volume of air completely stopped, divided by the mass of the air in the volume element.

Exposure assessment—Process of measuring or estimating concentration (or intensity), duration, and frequency of exposures to an agent present in the environment or, if estimating hypothetical exposures, that might arise from the release of a substance, or radionuclide, into the environment.

Exposure limit—General term defining an administrative substance concentration or intensity of exposure that should not be exceeded.

Exposure ratio—In a case control study, value obtained by dividing the rate at which persons in the case group are exposed to a risk factor (or to a protective factor) by the rate at which persons in the control group are exposed to the risk factor (or to the protective factor) of interest.

Exposure surface—Surface on a target where a substance (e.g., a pesticide) is present. With mammals, examples of outer exposure surfaces include the exterior of an eyeball, the skin surface, and a conceptual surface over the nose and open mouth. Examples of inner exposure surfaces include the gastrointestinal tract, the respiratory tract, and the urinary tract lining.

Exposure test—Determination of the level, concentration, or uptake of a potentially toxic compound and (or) its metabolite(s) in biological samples from an organism (blood, urine, hair, etc.) and the interpretation of the results to estimate the absorbed dose or degree of environmental pollution; or the measuring of biochemical effects, usually not direct adverse effects of the substance, and relating them to the quantity of substance absorbed, or to its concentration in the environment.

Food additive—Any substance, not normally consumed as a food by itself and not normally used as a typical ingredient of a given food, whether or not it has nutritive value, that is added intentionally to food for a technological (including organoleptic) purpose in the manufacture, processing, preparation, treatment, packing, packaging, transport, or holding of the food. Addition results, or may be reasonably expected to result (directly or indirectly), in the substance or its byproducts becoming a component of, or otherwise affecting, the characteristics of the food to which it is added. Note: The term does not include "contaminants" or substances added to food for maintaining or improving nutritional qualities.

Genetic toxicology—Study of chemically or physically induced changes to the structure of DNA, including epigenetic phenomena or mutations that may or may not be heritable.

Genotoxic—Capable of causing a change to the structure of the genome.

Good laboratory practice (GLP) (principles)—Fundamental rules incorporated in OECD guidelines and national regulations concerned with the process of effective organization and the conditions under which laboratory studies are properly planned, performed, monitored, recorded, and reported.

Guideline for exposure limits—Scientifically judged quantitative value (a concentration or number) of an environmental constituent that ensures esthetically pleasing air, water, or food and from which no adverse effect is expected concerning noncarcinogenic endpoints, or that gives an acceptably low estimate of lifetime cancer risk from those substances which are proven human carcinogens or carcinogens with at least limited evidence of human carcinogenicity.

Guideline value—Quantitative measure (a concentration or a number) of a constituent of an environmental medium that ensures esthetically pleasing air, water, or food and does not result in a significant risk to the user.

Harmful occupational factor—Component of the work environment, the effect of which on a worker under certain conditions leads to ill health or reduction of working ability.

Harmful substance—Substance that, following contact with an organism, can cause ill health or adverse effects either at the time of exposure or later in the life of the present and future generations.

Hazard—Set of inherent properties of a substance, mixture of substances, or a process involving substances that, under production, usage, or disposal conditions, make it capable of causing adverse effects to organisms or the environment, depending on the degree of exposure; in other words, it is a source of danger. See also risk.

Hazard assessment—Determination of factors controlling the likely effects of a hazard such as the dose—effect and dose—response relationships, variations in target susceptibility, and mechanism of toxicity.

Hazard evaluation—Establishment of a qualitative or quantitative relationship between hazard and benefit, involving the complex process of determining the significance of the identified hazard and balancing this against identifiable benefit. Note: This may subsequently be developed into a risk evaluation.

Hazard identification—Determination of substances of concern, their adverse effects, target populations, and conditions of exposure, taking into account toxicity data and knowledge of effects on human health, other organisms, and their environment.

In silico—Phrase applied to data generated and analyzed using computer modeling and information technology.

In vitro—In glass, referring to a study in the laboratory usually involving isolated organ, tissue, cell, or biochemical systems.

In vivo—In the living body, referring to a study performed on a living organism.

Inhibitory concentration (IC)—Concentration of a substance that causes a defined inhibition of a given system. Note: IC_{50} is the median concentration that causes 50% inhibition.

Inhibitory dose (ID)—Dose of a substance that causes a defined inhibition of a given system. Note: ID_{50} is the median dose that causes 50% inhibition.

Intake—Amount of a substance that is taken into the body, regardless of whether or not it is absorbed: The total daily intake is the sum of the daily intake by an individual from food, drinking-water, and inhaled air.

Intoxication—(1) Poisoning: pathological process with clinical signs and symptoms caused by a substance of exogenous or endogenous origin. (2) Drunkenness following consumption of beverages containing ethanol or other compounds affecting the central nervous system.

Lethal concentration (LC)—Concentration of a substance in an environmental medium that causes death following a certain period of exposure.

Lethal dose (LD)—Amount of a substance or physical agent (e.g., radiation) that causes death when taken into the body.

Lowest-effective dose (LED)—Lowest dose of a chemical inducing a specified effect in a specified fraction of exposed individuals.

Lowest-observed-adverse-effect level (LOAEL)—Lowest concentration or amount of a substance (dose), found by experiment or observation, that causes an adverse effect on morphology, functional capacity, growth, development, or life span of a target organism distinguishable from normal (control) organisms of the same species and strain under defined conditions of exposure.

Lowest-observed-effect level (LOEL)—Lowest concentration or amount of a substance (dose), found by experiment or observation, that causes any alteration in morphology, functional capacity, growth, development, or life span of target organisms distinguishable from normal (control) organisms of the same species and strain under the same defined conditions of exposure.

Maximum permissible level (MPL)—Level, usually a combination of time and concentration, beyond which any exposure of humans to a chemical or physical agent in their immediate environment is unsafe.

Maximum residue limit (MRL) (for pesticide residues)—Maximum contents of a pesticide residue (expressed as $mg\ kg^{-1}$ fresh weight) recommended by the Codex Alimentarius Commission[1] to be legally permitted in or on food commodities and animal feeds. Note: MRLs are based on data obtained following good agricultural practice and foods derived from commodities that comply with the respective MRLs are intended to be toxicologically acceptable.

Maximum residue limit (MRL) (for veterinary drugs)—Maximum contents of a drug residue (expressed as $mg\ kg^{-1}$ or $\mu g\ kg^{-1}$ of fresh weight) recommended by the Codex Alimentarius Commission to be legally permitted or recognized as acceptable in or on food commodities and animal feeds. Note: The MRL is based on the type and amount of residue considered to be without any toxicological hazard for human health as expressed by the acceptable daily intake (ADI) or on the basis of a temporary ADI that uses an additional uncertainty factor. It also takes into account other relevant public health risks as well as food technological aspects.

Maximum tolerable concentration (MTC)—Highest concentration of a substance in an environmental medium that does not cause death of test organisms or species (denoted by LC_0).

Maximum tolerable dose (MTD)—Highest amount of a substance that, when introduced into the body, does not kill test animals (denoted by LD_0).

---

[1] http://www.fao.org/fao-who-codexalimentarius/committees/cac/about/en/

Maximum tolerable exposure level (MTEL)—Maximum amount (dose) or concentration of a substance to which an organism can be exposed without leading to an adverse effect after prolonged exposure time.

Maximum tolerated dose (MTD)—High dose used in chronic toxicity testing that is expected on the basis of an adequate subchronic study to produce limited toxicity when administered for the duration of the test period. Note 1: It should not induce (1) overt toxicity, for example appreciable death of cells or organ dysfunction, or (2) toxic manifestations that are predicted materially to reduce the life span of the animals except as the result of neoplastic development, or (3) 10% or greater retardation of body weight gain as compared with control animals.

Median effective dose (ED_{50})—Statistically derived median dose of a chemical or physical agent (radiation) expected to produce a certain effect in 50% of test organisms in a given population or to produce a half-maximal effect in a biological system under a defined set of conditions. Note: ED_n refers to the median dose that is effective in n% of the test population.

Median lethal concentration (LC_{50})—Statistically derived median concentration of a substance in an environmental medium expected to kill 50% of organisms in a given population under a defined set of conditions.

Median lethal dose (LD_{50})—Statistically derived median dose of a chemical or physical agent (radiation) expected to kill 50% of organisms in a given population under a defined set of conditions.

Median lethal time (TL_{50})—Statistically derived median time interval during which 50% of a given population may be expected to die following acute administration of a chemical or physical agent (radiation) at a given concentration under a defined set of conditions.

Metabolic transformation—Biotransformation of a substance that takes place within a living organism.

Metabolism— (1) Sum total of all physical and chemical processes that take place within an organism from uptake to elimination. (2). In a narrower sense, the physical and chemical changes that take place in a substance within an organism, including biotransformation to metabolites.

Metabolite—Intermediate or product resulting from metabolism.

Minimum lethal concentration (LC_{min})—Lowest concentration of a toxic substance in an environmental medium that kills individual organisms or test species under a defined set of conditions.

Minimum lethal dose (LD_{min})—Lowest amount of a substance that, when introduced into the body, may cause death to individual species of test animals under a defined set of conditions.

Monitoring—Continuous or repeated observation, measurement, and evaluation of health and (or) environmental or technical data for defined purposes, according to prearranged schedules in space and time, using comparable methods for sensing and data collection. Note: Evaluation requires comparison with appropriate reference values based on knowledge of the probable relationship between ambient exposure and adverse effects.

Monte Carlo simulation—Analysis of a sequence of events using random numbers to generate possible outcomes in an iterative process.

Mortality study—Investigation dealing with death rates or proportion of deaths attributed to specific causes as a measure of response.

Multiple chemical sensitivity (MCS)—Intolerance condition attributed to extreme sensitivity to various environmental chemicals, found in air, food, water, building materials, or fabrics. Note: This syndrome is characterized by the patient's belief that his or her symptoms are caused by very low-level exposure to environmental chemicals. The term "chemical" is used to refer broadly to many natural and man-made chemical agents, some of which have several chemical constituents. Several theories have been advanced to explain the cause of multiple chemical sensitivity, including allergy, toxic effects, and neurobiological sensitization. There is insufficient scientific evidence to confirm a relationship between any of these possible causes and symptoms.

Multiple (or multiphasic) screening—Procedure that has evolved by combining single screening tests, and is the logical corollary of mass screening. Note 1: Where much time and effort have been spent by a population in attending for a single test such as mass radiography, it is natural to consider the economy of offering other tests at the same time. Note 2: Multiple (or multiphasic) screening implies the administration of a number of tests, in combination, to large groups of people.

Mutagenicity—Ability of a physical, chemical, or biological agent to induce (or generate) heritable changes (mutations) in the genotype in a cell as a consequence of alterations or loss of genes or chromosomes (or parts thereof).

Mycotoxin—Toxin produced by a fungus.

Nanotoxicology—Scientific discipline involving the study of the actual or potential danger presented by the harmful effects of nanoparticles on living organisms and ecosystems, of the relationship of such harmful effects to exposure, and of the mechanisms of action, diagnosis, prevention, and treatment of intoxications.

No-effect dose (NED)—Amount of a substance that has no effect on the organism. Note: It is lower than the threshold of harmful effect and is estimated while establishing the threshold of harmful effect.

No-effect level (NEL)—Maximum dose (of a substance) that produces no detectable changes under defined conditions of exposure. Note: This term tends to be substituted by no-observed-adverse-effect level (NOAEL) or no-observed-effect level (NOEL).

No-observed-adverse-effect level (NOAEL)—Greatest concentration or amount of a substance, found by experiment or observation, which causes no detectable adverse alteration of morphology, functional capacity, growth, development, or life span of the target organism under defined conditions of exposure.

No-observed-effect level (NOEL)—Greatest concentration or amount of a substance, found by experiment or observation, that causes no alterations of morphology, functional capacity, growth, development, or life span of target organisms distinguishable from those observed in normal (control) organisms of the same species and strain under the same defined conditions of exposure.

No-response level (NRL)—Maximum dose of a substance at which no specified response is observed in a defined population and under defined conditions of exposure.

Occupational exposure—Experience of substances, intensities of radiation, etc., or other conditions while at work.

Occupational exposure limit (OEL)—Regulatory level of exposure to substances, intensities of radiation, etc., or other conditions, specified appropriately in relevant government legislation or related codes of practice.

Occupational exposure standard (OES)—Level of exposure to substances, intensities of radiation, etc., or other conditions considered to represent specified good practice and a realistic criterion for the control of exposure by appropriate plant design, engineering controls, and, if necessary, the addition and use of personal protective clothing.

Occupational hygiene—Identification, assessment, and control of physicochemical and biological factors in the workplace that may affect the health or well-being of those at work and in the surrounding community.

Octan-1-ol–water partition coefficient, P_{ow}, K_{ow}—Ratio of the solubility of a chemical in octan-1-ol divided by its solubility in water. Note: Measure of lipophilicity, used in the assessment of both the uptake and physiological distribution of organic chemicals and prediction of their environmental fate.

Organic carbon partition coefficient, K_{oc}—Measure of the tendency for organic substances to be adsorbed by soil or sediment, expressed by the Eq. (3.6):

$$K_{oc} = \frac{\text{(mass adsorbed substance)}}{\text{(mass organic carbon)(mass concentration of adsorbed substance)}} \quad (3.6)$$

The K_{oc} is substance-specific and is largely independent of soil properties.

Partition coefficient—Concentration of a substance in one phase divided by the concentration of the substance in the other phase when the heterogeneous system of two phases is in equilibrium. Note 1: The ratio of concentrations (or, strictly speaking, activities) of the same molecular species in the two phases is constant at constant temperature. Note 2: The octan-1-ol–water partition coefficient is often used as a measure of the BCF for modeling purposes. Note 3: This term is in common usage in toxicology but is not recommended by IUPAC for use in chemistry and should not be used as a synonym for partition constant, partition ratio, or distribution ratio.

Partition ratio, K_D—Ratio of the concentration of a substance in a single definite form, A, in the extract to its concentration in the same form in the other phase at equilibrium, for example, for an aqueous/organic system, according the Eq. (3.7):

$$K_D(A) = \frac{[A]_{org}}{[A]_{aq}} \quad (3.7)$$

Permeability—Ability or power to enter or pass through a cell membrane.

Permeability coefficient, P—Quantity defining the permeability of molecules across a cell membrane and expressed as:

$$P = \frac{KD}{\Delta_x} \quad (3.8)$$

where K is the partition coefficient, D is the diffusion coefficient, and Δ_x is the thickness of the cell membrane. Note: SI units; m s^{-1}; frequently used units cm s^{-1}, with units cm^2 s^{-1} for D, cm for Δ_X.

Permissible exposure limit (PEL)—Recommendation by U.S. OSHA[2] for a time-weighted average concentration that must not be exceeded during any 8-hours work shift of a 40-hours working week.

---

[2] https://www.osha.gov/

Persistence—Attribute of a substance that describes the length of time that the substance remains in a particular environment before it is physically removed or chemically or biologically transformed.

Persistent inorganic pollutant (PIP)—Inorganic substance that is stable in the environment, is liable to long-range transport, may bioaccumulate in human and animal tissue, and may have significant impacts on human health and the environment. Note 1: Examples are arsenides, fluorides, cadmium salts, and lead salts. Note 2: Some inorganic chemicals, like crocidolite asbestos, are persistent in almost all circumstances, but others, like metal sulfides, are persistent only in unreactive environments; sulfides can generate hydrogen sulfide in a reducing environment or sulfates and sulfuric acid in oxidizing environments. As with organic substances, persistence is often a function of environmental properties.

Persistent organic pollutant (POP)—Organic chemical that is stable in the environment, is liable to long-range transport, may bioaccumulate in human and animal tissue, and may have significant impacts on human health and the environment. Examples: dioxin, PCBs, DDT, tributyltin oxide (TBTO). Note: The Stockholm Convention on Persistent Organic Pollutants was adopted at a Conference of Plenipotentiaries held from May 22–23, 2001 in Stockholm, Sweden; by signing this convention, governments have agreed to take measures to eliminate or reduce the release of POPs into the environment.

Personal monitoring—Type of environmental monitoring in which an individual's exposure to a substance is measured and evaluated. Note: This is normally carried out using a personal sampler.

Personal sampler—Compact, portable instrument for individual air sampling, measuring, or both, the content of a harmful substance in the respiration zone of a working person.

Pesticide residue—Any substance or mixture of substances found in humans or animals or in food and water following use of a pesticide: the term includes any specified derivatives, such as degradation and conversion products, metabolites, reaction products, and impurities considered to be of toxicological significance.

Poison—Substance that, taken into or formed within the organism, impairs the health of the organism and may kill it.

Poisoning intoxication—Morbid condition produced by a poison.

Pollutant—Any undesirable solid, liquid, or gaseous matter in a solid, liquid, or gaseous environmental medium. Note 1: "Undesirability" is often concentration-dependent, low concentrations of most substances being tolerable or even essential in many cases. Note 2: A primary pollutant is one emitted into the atmosphere, water, sediments, or soil from an identifiable source. Note 3: A secondary pollutant is a pollutant formed by chemical reaction in the atmosphere, water, sediments, or soil.

Quantitative structure—activity relationship (QSAR)—Quantitative structure—biological activity model derived using regression analysis and containing as parameters physicochemical constants, indicator variables, or theoretically calculated values. Note: The term is extended by some authors to include chemical reactivity, that is, activity and reactivity are regarded as synonyms. This extension is discouraged.

Quantitative structure—metabolism relationship (QSMR)—Quantitative association between the physicochemical and (or) the structural properties of a substance and its metabolic behavior.

Recommended exposure level (REL)—Highest allowable regulatory airborne concentration. Note: This exposure concentration is not expected to injure workers. It may be expressed as a ceiling limit or as a time-weighted average (TWA).

Reference dose (RfD)—An estimate (with uncertainty spanning perhaps an order of magnitude) of a daily oral exposure to the human population (including sensitive subgroups) that is likely to be without an appreciable risk of deleterious effects during a lifetime. Note: It can be derived from a NOAEL, LOAEL, or benchmark dose, with uncertainty factors generally applied to reflect limitations of the data used. It is generally used in USEPA's[3] noncancer health assessments.

Reference limit—Boundary value defined so that a stated fraction of the reference values is less than or exceeds that boundary value with a stated probability.

Residue—Contaminant remaining in an organism or in other material such as food or packaging, following exposure.

Response—Proportion of an exposed population with a defined effect or the proportion of a group of individuals that demonstrates a defined effect in a given time at a given dose rate.

Risk—(1) Probability of adverse effects caused under specified circumstances by an agent in an organism, a population, or an ecological system. (2) Probability of a hazard causing an adverse effect. (3) Expected frequency of occurrence of a harmful event arising from such an exposure.

---

[3] https://www.epa.gov/

Risk assessment—Identification and quantification of the risk resulting from a specific use or occurrence of a chemical or physical agent, taking into account possible harmful effects on individuals or populations exposed to the agent in the amount and manner proposed and all the possible routes of exposure. Note: Quantification ideally requires the establishment of dose—effect and dose—response relationships in likely target individuals and populations.

Risk characterization—Outcome of hazard identification and risk estimation applied to a specific use of a substance or occurrence of an environmental health hazard. Note: Risk characterization requires quantitative data on the exposure of organisms or people at risk in the specific situation. The end product is a quantitative statement about the proportion of organisms or people affected in a target population.

Risk evaluation—Establishment of a qualitative or quantitative relationship between risks and benefits, involving the complex process of determining the significance of the identified hazards and estimated risks to those organisms or people concerned with or affected by them.

Risk management—Decision-making process involving considerations of political, social, economic, and engineering factors with relevant risk assessments relating to a potential hazard so as to develop, analyze, and compare regulatory options and to select the optimal regulatory response for safety from that hazard. Note: Essentially risk management is the combination of three steps: risk evaluation; emission and exposure control; risk monitoring.

Risk monitoring—Process of following up the decisions and actions within risk management in order to check whether the aims of reduced exposure and risk are achieved.

Risk-specific dose—Amount of exposure corresponding to a specified level of risk.

Route of exposure—Means by which a toxic agent gains access to an organism by administration through the gastrointestinal tract (ingestion), lungs (inhalation), skin (topical), or by other routes such as intravenous, subcutaneous, intramuscular, or intraperitoneal routes.

Screening—(1) Carrying out of a test(s), examination(s), or procedure(s) in order to expose undetected abnormalities, unrecognized (incipient) diseases, or defects: Examples are mass X-rays and cervical smears. (2) Pharmacological or toxicological screening consists of a specified set of procedures to which a series of compounds is subjected to characterize pharmacological and toxicological properties and to establish dose—effect and dose—response relationships.

Screening level—Decision limit or cut-off point at which a screening test is regarded as positive.

Secondary metabolite—Product of biochemical processes other than the normal metabolic pathways, mostly produced in microorganisms or plants after the phase of active growth and under conditions of nutrient deficiency.

Sorption—Noncommittal term used instead of adsorption or absorption when it is difficult to discriminate experimentally between these two processes.

Standard(ized) morbidity ratio (SMR)—Ratio of the number of patients with a particular disease observed in a study group or population to the total number of people in the group or population multiplied by 100. Note: This ratio is usually expressed as a percentage.

Standard(ized) mortality ratio (SMR)—Ratio of the number of deaths observed in the study group or population to the number of deaths that would be expected if the study population had the same specific rates as the standard population, multiplied by 100. Note: This ratio is usually expressed as a percentage.

Structure—activity relationship (SAR)—Association between specific aspects of molecular structure and defined biological action. See also quantitative structure—activity relationship.

Structure—metabolism relationship (SMR)—Association between the physicochemical and (or) the structural properties of a substance and its metabolic behavior.

Susceptible vulnerable—Describing a group of organisms more vulnerable to a given exposure than the majority of the population to which they belong. Note: Susceptibility may reflect gender, age, physiological status, or genetic constitution of the organisms at risk.

Synergy—Pharmacological or toxicological interaction in which the combined biological effect of exposure to two or more substances is greater than expected on the basis of the simple summation of the effects of each of the individual substances.

Teratogenicity—(1) Potential to cause the production of non-heritable structural malformations or defects in offspring. (2) Production of nonheritable structural malformations or defects in offspring.

Testing of chemicals—(1) In toxicology, evaluation of the therapeutic and potentially toxic effects of substances by their application through relevant routes of exposure with appropriate organisms or biological systems so as to relate effects to dose following application. (2) In chemistry, qualitative or quantitative analysis by the application of one or more fixed methods and comparison of the results with established standards.

Three-dimensional quantitative structure–activity relationship (3D-QSAR)—Quantitative association between the three-dimensional structural properties of a substance and its biological properties.

Tolerable daily intake (TDI)—Estimate of the amount of a potentially harmful substance (e.g., contaminant) in food or drinking water that can be ingested daily over a lifetime without appreciable health risk. Note 1: For regulation of substances that cannot be easily avoided, a provisionally tolerable weekly intake (PTWI) may be applied as a temporary limit. Note 2: Acceptable daily intake is normally used for substances not known to be harmful, such as food additives.

Tolerance—(1) Adaptive state characterized by diminished effects of a particular dose of a substance: The process leading to tolerance is called "adaptation". (2) In food toxicology, dose that an individual can tolerate without showing an effect. (3) Ability to experience exposure to potentially harmful amounts of a substance without showing an adverse effect. (4) Ability of an organism to survive in the presence of a toxic substance. Increased tolerance may be acquired by adaptation to constant exposure. (5) In immunology, state of specific immunological unresponsiveness.

Total terminal residue (of a pesticide)—Summation of levels of all the residues of a defined pesticide in a food.

Toxicity—(1) Capacity to cause injury to a living organism defined with reference to the quantity of substance administered or absorbed, the way in which the substance is administered and distributed in time (single or repeated doses), the type and severity of injury, the time needed to produce the injury, the nature of the organism(s) affected, and other relevant conditions. (2) Adverse effects of a substance on a living organism defined as in 1. (3) Measure of incompatibility of a substance with life: This quantity may be expressed as the reciprocal of the absolute value of median lethal dose ($1/LD_{50}$) or concentration ($1/LC_{50}$).

Toxicity equivalency factor (TEF, f)—Ratio of the toxicity of a chemical to that of another structurally related chemical (or index compound) chosen as a reference.

Toxicity equivalency factor (TEF, f) (in risk assessment)—Ratio of the toxicity of a chemical to that of another structurally related chemical (or index compound) chosen as a reference. Factor used to estimate the toxicity of a complex mixture, commonly a mixture of chlorinated dibenzo-*p*-dioxins [oxanthrenes], furans, and biphenyls: In this case, TEF is based on relative toxicity to 2,3,7,8-tetrachlorodibenzo-*p*-dioxin [2,3,7,8-tetrachlorooxanthrene] for which the $f = 1$.

Toxicity equivalent (TEQ), Txe—Contribution of a specified component (or components) to the toxicity of a mixture of related substances. Note 1: The amount-of-substance (or substance) concentration of total toxicity equivalent is the sum of that for the components B, C ... N. Note 2: Toxicity equivalent is most commonly used in relation to the reference toxicant 2,3,7,8-tetrachlorodibenzo-*p*-dioxin [2,3,7,8-tetrachlorooxanthrene] by means of the toxicity equivalency factor (TEF, *f*) which is 1 for the reference substance. Hence, where c is the amount-of-substance concentration:

Toxic substance—Substance causing injury to living organisms as a result of physicochemical interactions.

Toxicity exposure ratio (TER)—Ratio of the measure of the effects (e.g., LD_{50}, LC_{50}, NOEC) to the estimated exposure. Note: It is the reciprocal of a risk quotient or hazard quotient.

Toxicity test—Experimental study of the adverse effects of exposure of a living organism to a substance for a defined duration under defined conditions.

Toxicokinetics—(1) Generally, the overall process of the absorption in biology (uptake) of potentially toxic substances by the body, the distribution of the substances and their metabolites in tissues and organs, their metabolism (biotransformation), and the elimination of the substances and their metabolites from the body. (2) In validating a toxicological study, the collection of toxicokinetic data, either as an integral component in the conduct of nonclinical toxicity studies or in specially designed supportive studies, in order to assess systemic exposure.

Toxicology—Scientific discipline involving the study of the actual or potential danger presented by the harmful effects of substances on living organisms and ecosystems, of the relationship of such harmful effects to exposure, and of the mechanisms of action, diagnosis, prevention, and treatment of intoxications.

Toxin—Poisonous substance produced by a biological organism such as a microbe, animal, plant, or fungus. Note: Examples are botulinum toxin, tetrodotoxin, pyrrolizidine alkaloids, and amanitin.

Validity of a measurement—Expression of the degree to which a measurement measures what it purports to measure.

Validity of a study—Degree to which the inferences drawn, especially generalizations extending beyond the study sample, are warranted when account is taken of the study methods, the representativeness of the study sample, and the nature of the population from which it is drawn.

We can observe either for ecotoxicology and toxicology a strong contribution of physicochemical, by means the proposition and understanding of several phenomena, as:
- Surface properties for sorption, with physical interaction and/or chemical bound presences
- Transport among phases and mediums, as permeation and partition
- Kinetics, with the velocity of processes and reactions
- Thermodynamics, with energetic aspects of reactions
- Others

They are directly related to the exposure to chemicals, its effects, and its fate. For instance, the quantitative structure-activity relationship (QSAR) is a predictive mathematical tool for (eco)toxicology based on physicochemical parameters determination, as K_{OW} and K_D, allied to biological parameters, as LC_{50} and LD_{50}.

Simple QSAR models calculate the toxicity of chemicals using a simple linear function of molecular descriptors presented in Eq. (3.9):

$$\text{Toxicity} = ax_1 + bx_2 + c \qquad (3.9)$$

Where x_1 and x_2 are the independent descriptor variables and a, b, and c are fitted parameters. The molecular weight and the octanol-water partition coefficient are examples of molecular descriptors (US Environmental Protection Agency, 2020a).

Taking LC_{50} and LD_{50} as parameters commonly used for the determination of toxicity, Table 3.1 describes toxicological data for pesticides and

Table 3.1 Toxicological data for some pesticides and veterinary drug which their residues can be determined in food and in the environment, according the scientific literature.

Parameter and route (mg kg^{-1})	Boscalid (fungicide)	Chlorpyrifos (insecticide)	Glyphosate (herbicide)	Imidacloprid (insecticide)	Oxytetracycline (antibiotic)
LD_{50} oral	>5000	82	4873	410	4800
LD_{50} dermal	>2000	202	7940	>5	—
LD_{50} inhalation		200	—	—	—

LD_{50}, the dose of the compound that causes mortality in 50% of treated population; LD, lethal dose.
Source: Published with permission from, Pubchem. (2020). <https://pubchem.ncbi.nlm.nih.gov/> Accessed 9.20.

Table 3.2 Ecotoxicological data for some pesticides and veterinary drug which their residues can be determined in food and in the environment, according the scientific literature.

Parameter and route (mg L^{-1})	Boscalid (fungicide)	Chlorpyrifos (insecticide)	Glyphosate (herbicide)	Imidacloprid (insecticide)	Oxytetracycline (antibiotic)
LC$_{50}$ oral	>5000	82	35–56	250–269	150

LC$_{50}$, the concentration of the compound to which the organisms were exposed that mortality in 50% of an exposed population; *LC*, lethal concentration.
Source: Published with permission from, Pubchem. (2020). <https://pubchem.ncbi.nlm.nih.gov/> Accessed 9.20.

veterinary drugs, whereas Table 3.2 describes ecotoxicological data for these same classes of agrochemicals (Pubchem, 2020).

LD$_{50}$ and LC$_{50}$ are helpful parameters to understand the effects caused by the chemical residue on the organisms in order to supply information to establish limits of exposure.

3.2 Disorders and diseases associated to agrochemical residues

Examples of diseases caused by chemical residues—or contaminants—in humans are presented in the Table 3.3.

It is expected that if the agrochemical residue achieve the necessary dose to produce the effect—the dose-response relationship—the second may occur, especially in the case of residues in food due to the easy sorption by the organism promoted by this exposure pathway.

In Europe the MRL for pesticide residues in food is set at the lowest limit of analytical detection (LOD). That is the MRL also for crops on which the pesticide has not been used or when its use has not left detectable residues. The default lowest limit (LOD) in EU law is 0.01 mg kg^{-1} (European Commission, 2020).

3.3 Dietary exposure

According the World Health Organization (2009), international dietary exposure assessments need to be performed for all identified chemical hazards present in the diet, and similar methods should be appropriate for:
- Contaminants
- Pesticide and veterinary drug residues

Table 3.3 Diseases and health effects associated to some agrochemical molecules.

Agrochemical molecule	Class	Diseases/health effects
Azithromycin	Veterinary drug (antibiotic)	Abdominal pain; acute kidney injury; depressive disorder; muscular diseases
Boscalid	Pesticide (fungicide)	Fatty liver; glucose intolerance; hyperglycemia; weight gain
Chlorpyrifos	Pesticide (acaricide and insecticide)	Anemia; anxiety disorders; bone narrow diseases; brain injuries
Glyphosate	Pesticide (herbicide)	Acute kidney injury; abnormalities; arthritis; rheumatoid
Oxytetracycline	Veterinary drug (antibiotic)	Acute kidney injury; bronchopneumonia; coma; confusion

Source: Adapted and published with permission from Pubchem (2020). <https://pubchem.ncbi.nlm.nih.gov/> Accessed 9.20.

- Nutrients
- Food additives (including flavorings)
- Processing aids and other chemicals in foods

Considering that, the dietary exposure can generate those effects presented in Table 3.3, there are a huge number of chemical species to be analyzed to evaluate their presence in food matrix, which is very heterogeneous according the Chapter 2, Agricultural matrices.

The general equation Eq. (3.10) for both acute and chronic dietary exposure is:

$$\text{Dietary exposure} = \sum (\text{concentration of chemical in food} \times \text{food consumption}) \text{body weight (kg)} \quad (3.10)$$

Thinking about the analysis to be carrying out, it is expected the application, for instance, of a chromatographic method to determine the organic chemical residue in food.

3.4 Occupational exposure

Mainly in the case of pesticide application, workers are frequently exposed to agrochemicals during their workday (Fig. 3.2), which promotes the

80 Analysis of Chemical Residues in Agriculture

Figure 3.2 Pesticide application in a rice planting. *Reproduced from internet (https:// www.portalvidanocampo.com.br/os-riscos-da-aplicacao-de-agrotoxicos-irregulares-na-producao/).*

occupational exposure. The intensity of this exposure and of their possible effects depends on:
- Characteristics of the application area, as presence of ventilation—air dilution helps to minimize the exposure effects
- Physicochemical properties of the AI and its additives, as vapor pressure, molecular weight, solubility, octanol-water partition coefficient (K_{OW}), etc.
- Routes of exposure, for example, inhalation, ingestion, and dermal (skin)
- Time of exposure
- Toxicity of the AI and its additives
- Use of personal protective equipment

García-García et al. (2016) compiled data related to the health effects of occupational exposure for greenhouse workers (Table 3.4). It is possible to observe a diversity of symptoms, as neurological, ocular, respiratory, digestive, skin, and osteomuscular disorders.

3.5 Environmental exposure

After the occupational exposure, the environmental exposure to agrochemicals is subject of concern due to the large use, for instance, of pesticides and fertilizers. Fig. 3.3 illustrates a pathway of exposure, fate, and associated effects.

Table 3.4 Human health effects attributable to pesticide exposure among greenhouse farm workers.

Clinical symptoms	Low pesticide exposure Controls 0–1 Symptom	≥ 2 Symptoms	Greenhouse workers 0–1 Symptom	≥ 2 Symptoms	P-value OR (95% CI)	High pesticide exposure Controls 0–1 Symptom	≥ 2 Symptoms	Greenhouse workers 0–1 Symptom	≥ 2 Symptoms	P-value OR (95% CI)
Neurological	42 (54.4%)	35 (45.5%)	59 (55.1%)	48 (44.9%)	$P=.936$ OR: 0.98 (0.54–1.75)	10 (30.3%)	23 (69.7%)	5 (26.3%)	14 (73.7%)	$P=.760$ OR: 1.22 (0.34–4.30)
Psychological	56 (71.8%)	22 (28.2%)	87 (81.3%)	20 (18.7%)	$P=.127$ OR: 0.58 (0.29–1.17)	23 (69.7%)	10 (30.3%)	17 (89.5%)	2 (10.5%)	$P=.172$ OR: 0.27 (0.05–1.39)
Ocular	54 (69.2%)	24 (30.8%)	68 (63.6%)	39 (36.4%)	$P=.421$ OR: 1.29 (0.69–2.40)	22 (66.7%)	11 (33.3%)	16 (78.9%)	4 (21.1%)	$P=.347$ OR: 0.53 (0.14–1.99)
Respiratory	62 (79.5%)	16 (20.5%)	78 (73.6%)	28 (26.4%)	$P=.354$ OR: 1.39 (0.69–2.79)	28 (84.8%)	5 (15.2%)	14 (73.7%)	5 (26.3%)	$P=.467$ OR: 2.00 (0.49–8.08)
Digestive	64 (82.1%)	14 (17.9%)	82 (76.6%)	25 (23.4%)	$P=.372$ OR: 1.39 (0.67–2.90)	28 (84.8%)	5 (15.2%)	14 (73.7%)	5 (26.3%)	$P=.467$ OR: 2.00 (0.49–8.08)
Skin	56 (71.8%)	22 (28.2%)	89 (83.2%)	18 (16.8%)	$P=.063$ OR: 0.51 (0.25–1.04)	15 (45.5%)	18 (54.5%)	9 (47.4%)	10 (52.6%)	$P=.894$ OR: 0.93 (0.30–2.87)
Osteomuscular disorders	77 (98.7%)	1 (1.3%)	76 (71.0%)	31 (29.0%)	$P=.000$ OR: 31.41 (4.18–235.9)	25 (75.8%)	8 (24.2%)	16 (84.2%)	3 (15.8%)	$P=.726$ OR: 0.59 (0.13–2.54)

Source: Reproduced with permission from Elsevier, García-García, C. R., Pasrón, T., Requena, M., Alarcón, R., Tsatsakis, A. M., & Hernández, A. F. (2016). Occupational pesticide exposure and adverse health effects at the clinical, hematological and biochemical level. Life Sciences, *145*, 274–283.

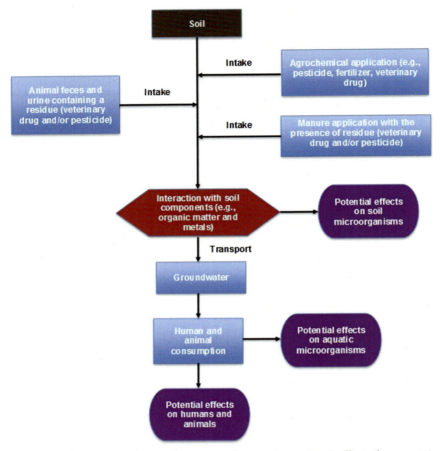

Figure 3.3 A generic pathway of exposure, fate, and associated effects for a certain agrochemical. *Credit: author.*

From the content of Fig. 3.3, we can note:
- Exposure: water consumption
- Fate: soil, water, organisms and microorganism
- Associated effects: antibiotic resistance, diseases, among others

According to the US Environmental Protection Agency (2020b), there are a set of information related to the environmental risk studies to be presented for a pesticide registration, highlighting the residue chemistry:
- Environmental fate, or what happens to the pesticide in soil, water, and air after being released into the environment.

- Non-target insect, or how the pesticide affects insects other than the ones the pesticide is intended to kill.
- Plant protection, or how the pesticide affects various plant species.
- *Residue chemistry*, or how much pesticide remains after application over time.
- Spray drift, or how much the pesticide drifts off-site when sprayed from the air. Helps determine exposure of non-target organisms.
- Wildlife and aquatic organisms, or how the pesticide affects various animal species.

An analytical study will supply data according the reliability required for the measures.

Moreover, it is very important take into account the toxicity of mixtures of residues due to a biochemical synergy promoting combined biological effect of exposure to two or more substances, as well as the potential toxic effects of agrochemicals and their metabolites on the natural microbiota (e.g., bacteria and fungi) of different matrixes. For instance, Rodea-Palomare, González-Pleiter, Martín-Betancor, and Fernández-Piñas (2015) observed the importance of pollutant mixtures in ecotoxicology in order to discuss the possibility of predicting interactions among mixture components at two levels: interactions among pairs of chemicals, and multicomponent interactions.

3.6 Conclusion

Ecotoxicological and toxicological scientific branches play a central role in the understanding of exposure, fate, and effects of agrochemicals and their residues for organisms and for the environment. Associated to their terms and usages, chemical analyzes are the tools to generate data and information to establish, for instance, exposure limits to certain chemical compounds—observing parameters as LD_{50} and LC_{50} and risk assessment—and to monitor its presence in food, soil, water, among other analytical matrixes.

References

Duffus, J. H., Norberg, M., & Templeton, D. M. (2009). Glossary of terms used in toxicology—2nd edition. IUPAC Recommendations 2007. *Pure and Applied Chemistry*, 79, 1153–1344.

European Comission. (2020). *How are EU MRLs set?* <https://ec.europa.eu/food/plant/pesticides/max_residue_levels/application_en> Accessed 9.20.

García-García, C. R., Pasrón, T., Requena, M., Alarcón, R., Tsatsaskis, A. M., & Hernández, A. F. (2016). Occupational pesticide exposure and adverse health effects at the clinical, hematological and biochemical level. *Life Sciences, 145*, 274–283.

Nordberg, M., Templeton, D. M., Andersen, O., & Duffus, J. H. (2009). Glossary of terms used in ecotoxicology. IUPAC Recommendations 2009. *Pure and Applied Chemistry, 81*, 829–970.

Pubchem. (2020). <https://pubchem.ncbi.nlm.nih.gov/> Accessed 9.20.

Rodea-Palomare, I., González-Pleiter, M., Martín-Betancor, K., & Fernández-Piñas, R. R. (2015). Additivity and interactions in ecotoxicity of pollutant mixtures: Some patterns, conclusions, and open questions. *Toxics, 3*, 342–369.

US Environmental Protection Agency. (2020a). Toxicity estimation software tool (TEST). <https://www.epa.gov/chemical-research/toxicity-estimation-software-tool-test> Accessed 9.20.

US Environmental Protection Agency. (2020b). Factsheet on ecological risk assessment for pesticides. <https://www.epa.gov/pesticide-science-and-assessing-pesticide-risks/fact-sheet-ecological-risk-assessment-pesticides> Accessed 9.20.

World Health Organization. (2009). *Principles and methods for the risk assessment of chemicals in food*. Geneva: World Health Organization.

CHAPTER 4

Fundamentals of analytical chemistry

The role of analytical chemistry in the analysis and understanding of the chemical residues from agriculture effects on humans and on the environment is paramount. The presence and effect of a certain compound are possible to be defined if analytical techniques and methods are applied, comprising: pollutant identification, exposure assessment, risk characterization, and legal regulation aspects (Pierzynski, Sims, & Vance, 2005).

Generally, chemical analysis can be considered as the use of concepts of analytical chemistry and its techniques and methods in the investigation and solution of real problems of variable complexity in different scientific or technological areas. The chemical analysis can generate information of both qualitative and quantitative character.

In order to understand the application of analytical techniques for the analysis of agrochemical residues in several matrixes, it is of fundamental importance to introduce some basic terms of analytical chemistry.

Initially, it should be taken into account that chemical analysis can be applied in three different or complementary situations:
- *Characterization*: observation of some physical property attributed to the *analyte*—the species of interest in the analytical process. For example, the absorption of visible radiation in the wavelength range of 400–450 nm or the behavior of the molecule against the incidence of radiation of other wavelengths—this is the typical application of certain spectroscopic techniques (infrared, nuclear magnetic resonance) and microscopic techniques.
- *Identification*: qualitative information on the presence or absence of the analyte—a good example is mass spectrometry, which identifies the compounds from the fragmentation of their molecular structure.
- *Determination*: quantitative information on the analyte concentration in the sample—an example is the elemental analysis of the composition and the chromatographic analyzes coupled to the detection techniques.

4.1 Analytical chemistry in the 21st century

Analytical chemistry, as a scientific branch of chemistry, is a generator of knowledge related to the characterization, identification, and determination for several materials from several origins in all states of matter.

Considering a broad context, analytical sciences are the experimental basis to understand the matter composition in completely different areas and economic sectors, in order to guarantee, for instance, the best uses of chemicals and related materials in the modern society:
- Advanced materials—for properties determination, for example, surface area, particle size, morphology.
- Agriculture—for assessment of the presence of agrochemical residues in agricultural products.
- Environment—for control and monitoring of pollutants in air, soil and water, and related matrixes.
- Industry—for quality control (QC) of raw materials, products, and processes.
- Life sciences—for diagnostic methods, for example, clinical analyzes.
- Others—for forensic, art and heritage investigation or evaluation.

Adams and Adriaens (2020) observed that there is a great number of fundamental works to do in ensuring the quality of data collection, data handling, and data reduction. Additionally, the use of chemometrics is necessary to transform data in actionable insight. Eventually, "smart metrology"—that is, artificial intelligence (AI)—will play a role in all this. However, the potential is remarkable as is the increasing use of AI in analytical chemistry,[1] which can be allied to advanced chemical instrumentation.

Moreover, aspects of sustainability—environmental, economic and societal impacts—related to the analytical processes should be taken into account when analytical chemistry and its concepts, techniques, and methods will be applied to solve real problems.

4.2 Figures of merit

Figures of merit are validation parameters used in analytical chemistry for the application of a certain analytical method. The most representative figures are *accuracy, linearity, limit of detection, limit of quantification, precision, selectivity, sensitivity,* and *robustness* (International Conference on Harmonization of Technical

---

[1] https://theanalyticalscientist.com/techniques-tools/do-androids-dream-of-analytical-chemistry

Requirements for Registration of Pharmaceuticals for Human Use, 2005). Before validating the analytical method, these parameters must be defined, as well as the limits at which results can be accepted.

4.2.1 Accuracy

Represents the degree of agreement between a measured value and a value taken as a "true value". The accuracy expresses the relative error of the measure Eq. (4.1):

$$AC = \frac{V_D - V_T}{V_D} \times 100 \tag{4.1}$$

Where: V_T is the true value and V_D is the determined value.

4.2.2 Linearity

Expresses the agreement between the results obtained by a given analytical method for a given parameter, such as the absorbance, and the analyte concentration, in a given concentration range. The linear correlation coefficient (r), calculated by the linear regression equation Eq. (4.2), is used to indicate if the mathematical model is adequate. Alternatively, one can use the coefficient of determination r^2, that the closer to 1 (one) the greater the linearity.

$$y_i = a + bx_i \tag{4.2}$$

Where: a is the line intercept and b is its slope coefficient.

4.2.3 Limit of detection and limit of quantification

The limit of detection (LOD) for an analytical method may vary depending on the type of sample and is defined as the minimum concentration of a measured and declared substance with 95% or 99% of confidence that the analyte concentration is greater than zero.

There are several ways to calculate LOD, but the recommendation is that at least seven replicates of the blank are made in the calculation. The quantification limit (LOQ) is the lowest analyte concentration that can be determined with an acceptable level of accuracy; can be considered as the mean value of the blank readings by adding 5, 6, or 10 times the standard deviation [Eq. (4.5) ahead]. The equations commonly used to determine LOD Eq. (4.3) and LOQ Eq. (4.4) are:

$$LOD = \frac{3.3s}{S} \tag{4.3}$$

$$\text{LOQ} = \frac{10s}{S} \quad (4.4)$$

Where: s is the standard deviation of the mean and S is the slope of the calibration curve [or b in Eq. (4.2)].

4.2.4 Precision

Degree of agreement between indications or measured values, obtained by repeated measurements, on the same object or similar objects, under specified conditions. It is generally expressed in numerical form by means of dispersion measures such as standard deviation, variance or coefficient of variation, under specified measurement conditions (recovery, repeatability or reproducibility).

The importance of precision Eq. (4.5) and its modes of measurement in analytical chemistry must be highlighted, which the main measurement form is the standard deviation. A measure of the data precision can be obtained by the population standard deviation (σ); or, more commonly, by the standard deviation of the mean (s) (International Standard Organization, 1993).

$$s = \sqrt{\frac{\sum (xi - \overline{x})}{n-1}} \quad (4.5)$$

Where: x_i is the value of a given measure; $\overline{x}$ is the arithmetic mean of the values of the measures $(\overline{x} = \sum xi/n)$; n is the number of measurements taken.

The coefficient of variation Eq. (4.6), or relative standard deviation, is useful to observe the relative accuracy of the measurements:

$$\text{CV}(\%) = \frac{s}{\overline{x}} \times 100 \quad (4.6)$$

The confidence interval for the mean Eq. (4.7) is very useful when expressing the confidence interval of a measure, a relevant aspect in the elaboration of an analytical report:

$$\text{CI}_M = \overline{x} \pm t_{n-1} \frac{s}{\sqrt{n}} \quad (4.7)$$

Where: $\overline{x}$ is the arithmetic mean of the values of the measures $(\overline{x} = \sum xi/n)$; n is the number of measurements performed; s is the standard deviation of the mean Eq. (4.5); t_{n-1} is the tabulated critical value of the student distribution (seen ahead).

4.2.5 Sensitivity or sensibility

It is the measure of the ability to discriminate between small differences in the concentration of an analyte Eq. (4.8). Two factors limit sensitivity: the slope of the analytical curve and reproducibility. For two methods having the same precision, the one with the most inclined analytical curve will be the most sensitive; if the analytical curves are equal, the one that exhibits greater precision will be more sensitive.

$$\Delta CA = \frac{\Delta S_A}{k_A} \quad (4.8)$$

Where: ΔS_A is the smallest increase in the signal that can be measured (the smallest difference is the analyte concentration that can be detected) and k_A is the proportionality constant to be measured.

4.2.6 Selectivity

The property of a measurement system, whereby the system provides measured values for one or several measurments (magnitude to be measured), such that the values of each measure are independent of each other. If the chosen method does not exhibit selectivity, the matrix components will interfere with the measurement performance. The evaluation of the selectivity of a method involves assays with reference standards or materials, samples with and without the analyte, and evaluation of the efficiency in determining the analyte in the presence of interferents—if S is equal to unity means that the method is selective for the analyte Eq. (4.9):

$$S = K_A \left(C_A + K_{A,I} C_I \right) \quad (4.9)$$

Where: K_A is the analyte's sensitivity coefficient (calculated from the Eq. (4.8)), C_A is the analyte concentration, $K_{A,I}$ is the selectivity coefficient, and C_I is the interfering concentration.

4.2.7 Robustness

Measures the sensitivity of a method to small variations in the conditions of analysis. A method is said to be robust when it is practically insensitive to such variations. Therefore, the greater the robustness, the greater the confidence of the method related to the precision. The coefficient of variation Eq. (4.6) can express this parameter.

4.2.8 Recovery

A figure of merit to be treated separately is the recovery percentage Eq. (4.10), which is important for determining the efficiency of an extraction method, with its value varying between 70% and 120%:

$$\%R = \frac{(C_i - C_f)}{C_i} \times 100 \qquad (4.10)$$

Where: C_i is the initial added concentration of the standard to the matrix, with no traces of the analyte; C_f is the final concentration determined in the sample (matrix + standard) after the addition of a known concentration of the standard, and after the application of an extraction method.

In any measure we carry out there are errors or uncertainties associated with it. The word *error* can be understood in two distinct ways: it can refer to the difference between a measured value and a known value, or related to the estimated uncertainty associated with a measurement or an experiment. Thus, the error can be classified as: random or indeterminate, systematic or determinate, and rough:

- Random errors exist in every measure, and cannot be totally eliminated, because they are caused by uncontrollable variables in the measurement process—these errors affect the precision of the results.
- Systematic errors have a defined cause, being of the same order of magnitude for replicates of a measure made in a similar way. They can be caused, for example, due to the lack of calibration of an equipment—these errors affect the accuracy of the results.
- Rough errors are usually of great magnitude, caused by human failure. These errors lead to anomalous values that differ significantly from the other replicated values, and there are several statistical tests to identify this type of error, such as the coefficient of variation.

4.3 Developing an analytical method

The stages of development and validation make up the *modus operandi* of any analytical method. As an example of the importance of these steps, we can mention the constant evaluation of the figures of merit (already seen) for the correct obtaining of an analytical result, a fundamental procedure in an accreditation process (seen ahead) of an analytical laboratory.

The development of a method for residue analysis requires, previously:
- Survey of the historic of the area and the agrochemical(s) application—involves consulting documents such as construction and

production plans, maps, soil, water or air data, and reports of previous analyzes.
- Toxicity of the analyte(s)—involves the understanding of the effect of the presence of the analyte(s) in the matrix—especially for environmental and food matrixes—and, in some cases, mathematical models of partition coefficients and mass transport estimation are used.
- Survey of the legal implication of the analyte(s) presence—search for knowledge about the maximum values allowed by the local and/or international legislation for the analyte(s) in the matrix to be analyzed, in order to know what should be the LOD and LOQ to be reached by the method.
- Survey of information in the technical-scientific literature—compilation of scientific and technical aspects obtained in articles, books and norms on the method to be developed.
- Survey of other information that is important, such as costs, logistics, and more adequate sampling.

For the development, adaptation, or implementation of a known method, an assessment process that evaluates the method efficiency in the routine of the laboratory should be applied. An essential step is to plan the activities to be performed so that the final result of the analysis performed is as reliable and representative as possible. In this sense, some observations are important:
- Calibrated equipment: the equipment and materials used in the analytical process should be properly calibrated.
- Quality of the analytical reagents: the laboratory needs reagents of high purity for analysis to avoid contaminant effects on the results. It is very critical when analyzing analytes in very small quantities or trace concentrations.
- Certified standards: whenever possible, we must work with certified standards, which contain uncertainty, and which are traceable.
- Calibrated glassware: glassware to be used in quantitative analyzes, such as pipettes, test tubes, etc., should be checked, observing the calibration temperature, in the case of the calibration process and estimated at a given temperature.
- Representative sampling: establish the correct way of sampling, according to the physicochemical characteristics of the material to be analyzed—a correct sampling ensures the reliability of the result.
- Statistical tools: check for correct data interpretation, for example, using software as Origin.[2]

---

[2] https://www.originlab.com/index.aspx?go = Products/Origin/Statistics

- Qualified personnel: analysts and technicians must be trained and qualified to perform the procedures, respecting their level of training.

In addition to these points addressed, it is essential that the method chosen attends in a satisfactory form to the analytical goal. In this respect, it is necessary to consider the sensitivity of the method, concentration, and matrix in which the analyte is present and a presence of chemical interferers, which may mask the result obtained;—it is of great importance to know the effect of the matrix on the result. Finally, once the method has been developed, it must be optimized so that it is then validated.

4.3.1 Calibration

It is often necessary for the analyst to provide references so that it is possible to correlate the data obtained in an apparatus with the real analyte concentration in the sample. This is done by means the construction of a calibration curve. This procedure is called calibration of the method, which is nothing more than to determine the relationship between the analytical response and the analyte concentration. The main methods used in the construction of the calibration curves are:

- *Standard addition*: addition of known concentrations of the analyte to known quantities of the sample to be analyzed, generating plots for the construction of the calibration curve from peak area and concentration values. By means the extrapolation of the curve on the abscissa axis the actual concentration of analyte in the sample is obtained; the difference between the results of the sample without addition and addition of analyte must be equal to the added concentration. The method can also be used with multiple additions of standard, which allows to verify if there is a linear relationship between the response and the concentration of the analyte. Normally, standard addition is used when the matrix has complex composition that affects the analytical signal, or when an analyte pattern cannot be found,—see Fig. 4.1A.
- *External standard*: when it is known that the constituents of the sample do not cause interference in the analyte signal, the external standard method can be used. The method consists in the construction of an analytical curve from the areas obtained with standard solutions of the analyte in known concentrations; special care must be taken with the preparation of the standard solutions, since any contamination will imply in an erroneous determination of the analyte concentration, see the Fig. 4.1B.

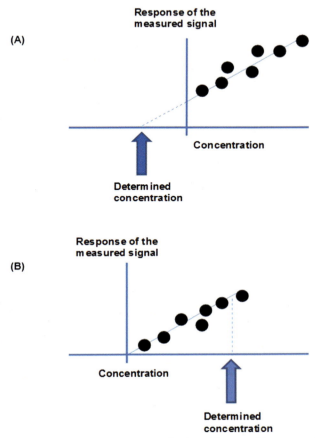

Figure 4.1 Calibration curves: (A) standard addition, (B) external standard. Credit: author.

- *Internal standard*: addition of a standard, which is a compound with a composition different from the analyte and of known concentration; but with a similar chemical structure allowing a behavior close to that of the analyte in relation to the analytical response to a series of analyte standards with known analyte concentrations for the construction of a calibration curve (Fig. 4.2). This curve is constructed not with the analyte response, but with the ratio of the internal standard signal to the analyte signal. Any analyte of unknown concentration can then be determined with addition of the internal standard by projection of the ratio of the responses in the analytical curve. This method is especially used when small variations in

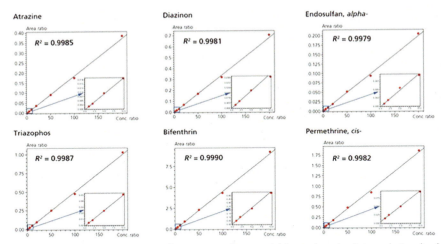

Figure 4.2 Calibration curves of representative pesticides using the internal standard method. *Courtesy Shimadzu (https://www.shimadzu.com/products/index.html).*

the response of the equipment to each analysis performed, as in chromatographic analysis, usually occurs.

Fig. 4.2 illustrates calibration curves for several pesticide residues for their analysis in botanical ingredients using gas chromatography with triple quadrupole mass spectrometry.

4.4 Validating an analytical method

Validation is proof by objective evidence that the requirements for a particular application or use of a method have been attended.

According to the International Standard Organization (2005) the laboratory shall validate non-standard methods, which are methods developed by the laboratory itself, or standard methods used outside the scope for which they were designed, such as extensions or modifications. The latter refers to methods developed by a standardization body or other segments whose methods are accepted by the technical sector concerned.

There are several definitions for validation in the literature. However, according Ribani, Bottoli, Collins, and Jardim (2004) the validation can proceed in two ways: validation in the laboratory and full validation. It is considered validation in the laboratory when it is used to verify the suitability of a method or when a method has been developed in the laboratory and all parameters are related to the measurements in that laboratory.

Fundamentals of analytical chemistry

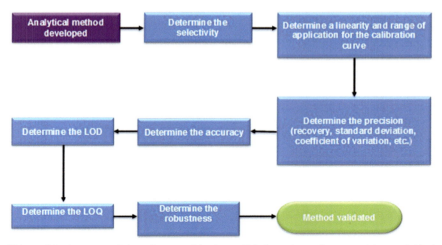

Figure 4.3 A sequential pathway with the validation steps for a certain analytical method. *Credit: author.*

Thus, validation in the laboratory is a preliminary step to full validation, which is performed considering all performance characteristics and interlaboratory tests. Fig. 4.3 illustrates the generic validation process of an analytical method.

For a method validation its results should be compared against results obtained from a *reference material*, which is defined as a "material, sufficiently homogeneous and stable with respect to one or more specified properties, which has been established to be fit for its intended use in a measurement process" (International Standard Organization, 2016). For a metrology purposes, a certified reference material (CRM) is defined as a "reference material characterized by a metrologically valid procedure for one or more specified properties, accompanied by a reference material certificate that provides the value of the specified property, its associated uncertainty, and a statement of metrological traceability" (International Standard Organization, 2016). As observed by Wise (2018), most CRMs are produced by National Metrology Institutes (NMIs) in a large number of countries, such as the National Institute of Standards and Technology (NIST)[3] in the United States.

---

[3] https://www.nist.gov/srm

4.4.1 Interlaboratory studies

The participation of the analytical laboratory in the interlaboratorial studies can verify if its adopted methodology is consolidated for a certain type of analysis, with a well-developed and defined control system.

The International Standard Organization (2005) recommends that a technique used to determine the interlaboratory performance of a method is as follows—or a combination of them:

- *Calibration with the use of materials or reference standards:* the reference material is a sufficiently homogeneous and stable material with respect to certain physical or chemical properties and is prepared to suit its intended use in a measurement or examination of qualitative properties and shall be accompanied by documentation issued by a notified body authority, a qualification or more property values specified as uncertainties and as associated traces, called in this case a CRM. Already the reference standard is used for the calibration of other standards of magnitude of the same type in a laboratory.
- *Comparisons with results obtained by other methods:* the efficiency of the developing method can be verified by comparing its results with the results of a standardized method, through statistical tests (International Standard Organization, 1994).

The most utilized test for the comparisons with results obtained by other methods is the student—or t-test, used to compare a mean of a series of results with a reference value, as means of two sets of results, within a confidence interval. The value found is compared with the tabulated value of t, and the first one should be as close as possible to the validation of the proposed method, according the Eq. (4.11):

$$t_{calc} = \frac{|\mu - \bar{x}|\sqrt{n}}{s} \qquad (4.11)$$

Where: $\bar{x}$ is the arithmetic mean for the set values; μ is the reference value, that can be substituted by the mean from another data set; n is the number of measurements; and s is the standard deviation of the mean. Table 4.1 presents t values according to the measurement number and the correctness probability for those measurements (Harvey, 2000).

Comparing the results of the two different methods or comparing the results of the two different laboratories can be done by means of the F-test, according Eq. (4.12):

$$F = \frac{sx^2}{sy^2} \qquad (4.12)$$

Table 4.1 Tabulated value of t, according the measurement number for 95% and 99% of confidence interval.

Degrees of freedom ($n - 1$)	Critical value of t for 95% of confidence interval	Critical value of t for 99% of confidence interval
1	12.71	63.66
2	4.30	9.93
3	3.18	5.84
4	2.78	4.60
5	2.57	4.03
6	2.45	3.71
7	2.37	3.50
8	2.31	3.36
9	2.26	3.25
10	2.23	3.17
∞	1.96	2.58

Source: Adapted from Harvey, D. (2000). Modern analytical chemistry. Boston: McGraw-Hill.

Where: s is the mean standard deviation for each measurement set (x or y).

The largest value of s is always used in the numerator, so the value of F will always be greater than the unit. The value found is then compared with the tabulated value of F, considering the degrees of freedom of each set of data. To be considered equally efficient the value found has to be less than the tabulated value. Table 4.2 presents F values for an exception probability of 5% of the cases (Miller & Miller, 2005).

4.4.2 Interlaboratory comparisons

The analysis of the same type of sample is carried out by several laboratories, whose objective is to verify if the result obtained by the laboratory that is developing the method is reproducible.

4.4.3 Systematic evaluation of factors influencing results

It is important that you have a good knowledge of the measurement process to evaluate the possible sources of interference in the final result and it should ideally be done continuously.

4.4.4 Evaluation of uncertainty of results generated

According to the International Vocabulary of Metrology, the uncertainty of a measurement is a parameter associated with the result that

Table 4.2 Tabulated values of F for 5% of probability of significance (P = .05)[a], according to the degrees of freedom of the numerator and denominator.

Degrees of freedom (denominator)	Degrees of freedom (numerator)						
	3	4	5	6	12	20	∞
3	9.28	9.12	9.01	8.94	8.74	8.64	8.53
4	6.59	6.39	6.26	6.16	5.91	5.80	5.63
5	5.41	5.19	5.05	4.95	4.68	4.56	4.36
6	4.76	4.53	4.39	4.28	4.00	3.87	3.67
12	3.49	3.26	3.11	3.00	2.69	2.54	2.30
20	3.10	2.87	2.71	2.60	2.28	2.12	1.84
∞	2.60	2.37	2.21	2.10	1.75	1.57	1.00

[a]The P-value is defined as the probability of the results of an experiment deviating from the null by as much as they did or greater if the null hypothesis is true. Traditionally, the cutoff value to reject the null hypothesis is .05, which means that, when there is no difference, a value as high as test statistic is expected to be less than of 5% of the time.
Source: Adapted from Miller, J. N., & Miller, J. C. (2005). Statistics and chemometrics for analytical chemistry (5th ed.). Harlow: Pearson.

characterizes the dispersion of the values obtained around the mean, since there are associated uncertainties in each measurement process (Bureau International des Poids et Mesures, 2012). The combined total uncertainty or standard uncertainty, u, is the sum of the uncertainties generated by the various components of the measurement process, each expressed as a standard deviation. Having established a confidence level, the expanded combined uncertainty, U, is determined by the confidence interval criterion, using a coverage factor, k. Most of the time, we use $k = 2$, corresponding to the confidence level of approximately 95% (Olivieri et al., 2006). The measurement of uncertainty should not be confused with the error. The error is defined as the difference between the measured value and the true value Eq. (4.13).

$$u = \frac{U}{k} \tag{4.13}$$

4.4.5 Repeatability and reproducibility

These two terms are frequently observed in metrology in chemistry[4] when we need to evaluate the reliability of analytical results for intra

---
[4] https://www.eurachem.org/index.php/cat-wg/wg-inact/wg-metchem

or interlaboratory studies for QC and quality assurance (QA) (item 4.6 ahead).

According to the International Union of Pure and Applied Chemistry (2020), we can define:
- Repeatability: the closeness of agreement between independent results obtained with the same method on identical test material, under the same conditions (same operator, same apparatus, same laboratory and after short intervals of time). The measure of repeatability is the standard deviation qualified as repeatability standard deviation.
- Reproducibility: the closeness of agreement between independent results obtained with the same method on identical test material but under different conditions (different operators, different apparatus, different laboratories, and/or after different intervals of time). The measure of reproducibility is the standard deviation qualified as reproducibility standard deviation.

4.4.6 Accreditation of an analytical laboratory for residue analysis

Accreditation provides independent confirmation of competence. For an analytical laboratory, it aims to guarantee the reliability of its results issued against quality parameters established and evaluated by a recognized accrediting body. It is required for an agency or official body to accept the laboratory results.

At present, accreditation is one of the main requirements for the performance of an analytical laboratory, taking into account the fact that it accredits the quality, since the accredited laboratory complies with the norm ISO/IEC 17025 (International Standard Organization, 2005). There are cases that Good Laboratory Practices (GLPs) (Organization for Economic Co-operation and Development, 1998) should also be considered in the accreditation process—such as for studies on the release of pesticides. Therefore, accreditation according to ISO/IEC 17025 (International Standard Organization, 2005) should be the motto of the quality of laboratories carrying out residue analyzes.

Note that accreditation takes place for a pre-defined analytical scope. That is, the fact that the laboratory is accredited in a certain herbicide analyzes in water does not mean that, for example, it is accredited to perform analyzes of organophosphorus pesticides in the same matrix; if not, it should request new accreditation to attend the second case. More information about accreditation process can obtained on the website of the

International Accreditation Forum (International Accreditation Forum, 2020).

In a general way, an accreditation process which can be applied to several analytical laboratories comprises the follow steps:
1. Accreditation requesting to the accrediting organism;
2. Prior analysis of the laboratory documentation by the organism;
3. If required documentation is in accordance, opening the accreditation process;
4. Verification of the relevance of the accreditation requesting;
5. Interlaboratory comparison for performance evaluation in the intended analytical scope;
6. If performance is satisfactory, analysis of the accreditation process by the Accreditation Committee;
7. Final decision by the Steering Committee (SC);
8. If the SC's final decision is favorable, signing of the Accreditation Agreement.

4.5 Chemometrics

In some applications, an analytical methodology alone is not sufficient to provide qualitative or quantitative information of the sample, using only data such as the intensity of absorption or emission, and/or the region of absorption of the electromagnetic spectrum—called univariate analysis. Often, the analysis is associated with chemometric tools to provide the best information.

Chemometrics can be understood as an area of knowledge of chemistry that uses mathematic models, along with formal logic to interpret and predict data, thus extracting the maximum of relevant information. It is largely used for spectroscopic and chromatographic data. In the case of spectroscopic data, each wavelength is a variable. Spectra or complete chromatograms, parts of them or selection of variables can be used. Since several variables are treated at the same time, the data analysis is called multivariate analysis. In order to carry out the multivariate analysis, the data are first organized in matrix form, called matrix **X** of original data, where the columns correspond to the predictor variables (such as absorbance) and the lines correspond, for example, to the concentration of an analyte (Martens & Naes, 1989).

After organizing the data in the matrix, sometimes it is necessary to pre-process, eliminating irrelevant information or standardizing the data.

The objective of the multivariate analysis can be from an exploratory analysis to the quantification of an analyte (Brereton, 2003). The exploratory analysis is performed with the objective of obtaining initial information from a set of samples, such as the formation of clusters according to a certain chemical property. The main chemometric tool used in the exploratory analysis is the PCA (principal component analysis). When it is desired to verify similarities between samples of a certain class, samples are classified, with the most common methods being KNN (k-nearest neighbor), LDA (linear discriminant analysis), HCA (hierarchical cluster analysis), SIMCA (soft independent modeling of class analogy). When it is intended to predict analyte concentration, calibration models are constructed, with patterns of known concentration and working range that contemplate the analyte concentration. The most widely used method for this purpose is PLS (partial least squares). For instance, Tables 4.3 and 4.4 illustrate the application of PCA on analytical data. Initially, Table 4.3 presents data for a flourescence hypothetical analysis.

From the data of Table 4.3 we obtain the covariance matrix—a joint variance[5] of two variables—in the Table 4.4.

This shows that, for example, the covariance for the fluorescence intensities at 350 and 400 nm is -1.15909. The table also gives the variances of the fluorescence intensities at each wavelength along the leading diagonal of the matrix: for the fluorescence intensities at 350 nm the variance is 2.75. We can consider this kind of information in a practical way to understand the propagation of errors, and consequent reliability, for a certain analysis.

Chemometrics application for quantitative and qualitative purposes can be generates, according Szymánka et al. (2015):
- *For qualitative results*: compound identification, compound classification, and sample classification.
- *For quantitative results*: sample calibration and QSAR (quantitative structure—activity relationship).

Chemometrics does not only apply to measurements, but also to the extraction step. Because it is based on multiparametric analyzes, it allows to evaluate the effect of the variation of the operational parameters on the recovery percentage values of the extraction method. It is possible, for example, to verify among several extraction methods the most suitable for a group of analytes; or the effect of the matrix on the analyte group

---

[5] Variance is the square of the standard deviation (s); covariance is the sum of variance for a certain measurement.

Table 4.3 Relative intensities of fluorescence emission at four different wavelengths (300, 350, 400, and 450 nm) for 12 compounds, A–L.

Compound	Wavelength (nm)			
	300	350	400	450
A	16	62	67	27
B	15	60	69	31
C	14	59	68	31
D	15	61	71	31
E	14	60	70	30
F	14	59	69	30
G	17	63	68	29
H	16	62	69	28
I	15	60	72	30
J	17	63	69	27
K	18	62	68	28
L	18	64	67	29
Mean	15.75	61.25	68.92	29.25
Standard deviation	1.485	1.658	1.505	1.485

Source: Adapted from Miller, J. N., & Miller, J. C. (2005). *Statistics and chemometrics for analytical chemistry* (5th ed.). Harlow: Pearson.

against more than one extraction method—it is very useful in environmental and food analysis.

Farina, Abdullah, Bibi, and Khalik (2017) applied a chemometric study to optimize the extraction in vegetable samples for screening and quantification of 15 pesticide residues at trace levels in cabbage, broccoli, cauliflower, lettuce, celery, spinach, and mustard. The chemometric method consists of two steps, first, to determine the significance of each factor by Pareto chart[6] followed by optimization of these significant factors using central composite design (CCD)[7] (Fig. 4.4). Minitab statistical software

---

[6] The Pareto diagram is a column chart that orders the frequencies of occurrences, from highest to lowest, allowing the prioritization of problems, trying to carry out the Pareto principle, that is, there are many minor problems in face of more serious ones.

[7] A Box-Wilson Central Composite Design, commonly called 'a central composite design' (CCD)—very useful for surface-response experimental design—contains an imbedded factorial or fractional factorial design with center points that is augmented with a group of 'star points' that allow estimation of curvature. If the distance from the center of the design space to a factorial point is ±1 unit for each factor, the distance from the center of the design space to a star point is $|\alpha| > 1$. The precise value of α depends on certain properties desired for the design and on the number of factors involved: https://www.itl.nist.gov/div898/handbook/pri/section3/pri3361.htm

Table 4.4 Covariance matrix for the data from Table 4.3.

λ (nm)	λ (nm)			
	300	350	400	450
300	2.20455			
350	2.25000	2.75000		
400	−1.11364	−1.15909	2.26515	
450	−1.47727	−1.70455	1.02273	2.20455

Source: Adapted from Miller, J. N., & Miller, J. C. (2005). *Statistics and chemometrics for analytical chemistry* (5th ed.). Harlow: Pearson.

was used for these multivariate experiments for the generation of 2^{4-1} design and CCD matrices.

4.6 Quality control and quality assurance

After development and validation, the analytical method requires permanent control. It is of extreme relevance, since it allows verifying the need for a revalidation. In general, a revalidation must be carried out, with one of the following situations (International Standard Organization, 1994):

- Introduction of a new analytical method in place of the previously validated one.
- Exchange of a particular reagent for another of different brand that has lower purity and quality specifications.
- Preventive or corrective maintenance in an instrument used in the methodology, altering the original technical configurations of the manufacturer.
- Changes in work concentration of the analytical method, and changes not predicted in parameters of the analytical method in the original robustness test.

The use of the previously described figures of merit (item 4.2) is essential for such methodological control.

According the US Environmental Protection Agency (2020), QA/QC measures are those activities that are undertaken to demonstrate the accuracy (how close to the real result) and precision (how reproducible results are)—These figures of merit were seen in the item 4.2. QA generally refers to a broad plan for maintaining quality in all aspects of a program. This plan comprises proper documentation of all procedures, training of volunteers, study design, data management and analysis, and specific QC measures. QC consists of the steps to determine the validity of specific

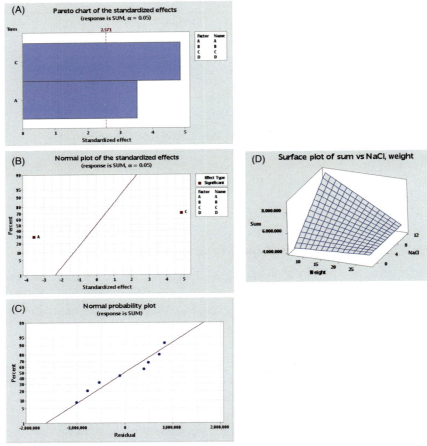

Figure 4.4 The multivariate optimization, where: (A) Pareto chart of standardized main effect; (B) normal plot of standardized main effect; (C) normal probability plot; and (D) surface plot of sum versus NaCl weight. *Reprinted with permission from Elsevier, Farina, Y., Abdullah, M. P., Bibi, N., & Khalik, W. M. A. W. M. (2017). Determination of pesticide residues in leafy vegetables at parts per billion levels by a chemometric study using GC-ECD in Cameron Highlands, Malaysia. Food Chemistry, 224, 55–61.*

sampling and analytical procedures with internal (e.g., field blank, lab replicates, spike samples, calibration blank and calibration standards) and external (e.g., internal field duplicates, split samples, knows and unknowns) checks. The assessment of the overall precision and accuracy of the generated data, after the analyses, is the quality assessment.

A quality management system is constituted by the followed components (Prichard & Barwick, 2007):
- Management structure and responsibility;
- Third-party assessment;
- Annual review (by senior management);
- Auditing (internal and external);
- Training (internal and external);
- Records (validation, calibration, QC, complaints);
- Documentation (central and local).

It is relevant to highlight the GLPs (Organization for Economic Co-operation and Development, 1998) as the reference of QA in studies involving agrochemical residues and the norm ISO/IEC 17025 (International Standard Organization, 2005) as the reference of competency requirements, according a quality management system, for analytical laboratories.

4.7 Green analytical chemistry

As discussed in the item 4.1, it is urgent to develop new techniques and methods according to sustainable approaches that promote a reduction in the negative impact from chemical processes on the public health and on the environment— in laboratory and in the industry. From this holistic view, the 12 fundamental principles of green chemistry (GC) can aid to achieve these premises. They comprise (ACS Green Chemistry Institute, 2020):

1. *Prevention*: it is better to prevent waste than to treat or clean up waste after it has been created.
2. *Atom economy*: synthetic methods should be designed to maximize the incorporation of all materials used in the process into the final product.
3. *Less hazardous chemical syntheses*: wherever practicable, synthetic methods should be designed to use and generate substances that possess little or no toxicity to human health and the environment.
4. *Designing safer chemicals*: chemical products should be designed to affect their desired function while minimizing their toxicity.
5. *Safer solvents and auxiliaries*: the use of auxiliary substances (e.g., solvents, separation agents, etc.) should be made unnecessary wherever possible and innocuous when used.
6. *Design for energy efficiency*: energy requirements of chemical processes should be recognized for their environmental and economic impacts

and should be minimized. If possible, synthetic methods should be conducted at ambient temperature and pressure.
7. *Use of renewable feedstocks*: a raw material or feedstock should be renewable rather than depleting whenever technically and economically practicable.
8. *Reduce derivatives*: unnecessary derivatization (use of blocking groups, protection/ deprotection, temporary modification of physical/chemical processes) should be minimized or avoided if possible, because such steps require additional reagents and can generate waste.
9. *Catalysis*: catalytic reagents (as selective as possible) are superior to stoichiometric reagents.
10. *Design for degradation*: chemical products should be designed so that at the end of their function they break down into innocuous degradation products and do not persist in the environment.
11. *Real-time analysis for pollution prevention*: analytical methodologies need to be further developed to allow for real-time, in-process monitoring, and control prior to the formation of hazardous substances.
12. *Inherently safer chemistry for accident prevention*: substances and the form of a substance used in a chemical process should be chosen to minimize the potential for chemical accidents, including releases, explosions, and fires.

Regarding to chemical analyzes, we can highlight principles 5, 8, 11, and 12 as those that prove to be the most interesting for application in a laboratory,—as discussed ahead.

The concept of green analytical chemistry can be addressed to these purposes (Armenta, Garrigues, & de la Guardia, 2008):
- Sample treatment;
- Oriented scanning methodologies;
- Alternatives to toxic reagents;
- Waste minimization;
- Recovery of reagents;
- The online decontamination of wastes;
- Reagent-free methodologies.

Waste prevention, safe solvents and auxiliaries, energy efficiency and inherently safer chemistry for accident prevention are obvious requirements for all chemical operations. Safer chemicals, the reduction of derivatives and the use of catalysts should be taken in account for each analysis because each analytical process has its own technical particularities. For example, the use of real-time analysis for pollution control is a good opportunity for technological development in analytical chemistry in the use of an in situ system for effluent analyses (gaseous and liquids). In a large number of cases is not possible to apply all of these principles due to the particularities of either the sample or the matrix, but is very important to consider these individually in an analytical process. This exercise will ensure the "greening" of the analysis.

As a practical guidance, De la Guardia and Garrigues (2011) established the main objectives to be considered as a green method:
- Simplication;
- The selection of reagents to be avoided based on toxicity, renewability or degradability data;
- The maximization of information;
- The minimization of consumables, taking into consideration the number of samples, the volumes or masses of reagents and energy consumption;
- The detoxication of wastes.

These objectives will define the best strategy to be applied as a result of the principles of GC. Furthermore, the application of GC principles already permeates toxicology for the studies of environmental pollutants, as demonstrated by Crawford et al. (2017), what can promote the reduction of animal testing.

4.8 Conclusion

The role of analytical chemistry in the analysis and understanding of the chemical residues from agriculture effects on humans and on the environment is paramount. To achieve this role, chemical analyzes should follow the figures of merit in order to guarantee the performance and reliability of a certain analytical method.

Developing and validating are required steps to assure the correct application of the analytical method in the agrochemical residues analyses. On the other hand, mathematic methods, especially chemometrics, offer the basis for data treatment and interpretation.

Furthermore, QA/QC and GC are aspects whose should be considered to explore, in a sustainable and concise ways, the best results provided by analytical chemistry.

References

ACS Green Chemistry Institute. (2020). *12 Principles of green chemistry*. <https://www.acs.org/content/acs/en/greenchemistry/what-is-green-chemistry/principles/12-principles-of-green-chemistry.html> Accessed 9.20.

Adams, F., & Adriaens, M. (2020). The metamorphosis of analytical chemistry. *Analytical and Bioanalytical Chemistry, 412*, 3525−3537.

Armenta, S., Garrigues, S., & de la Guardia, M. (2008). Green analytical chemistry. *TrAC, Trends in Analytical Chemistry, 27*, 497−511.

Brereton, R. G. (2003). *Chemometrics: Data analysis for the laboratory and chemical plant*. Chichester: John Wiley & Sons.

Bureau International des Poids et Mesures. BIPM. (2012). *International vocabulary of metrology—basic and general concepts and associated terms* (3rd ed.). Paris: BIPM.

Crawford, S. E., Hartung, T., Hollert, H., Mathes, B., van Ravenzwaay, B., Steger-Hartman, T., ... Krug, H. F. (2017). Green toxicology: A strategy for sustainable chemical and material development. *Environmental Sciences Europe, 29*, 1−16.

De la Guardia, M., & Garrigues, M. (Eds.), (2011). *Challenges in green chemistry*. Cambridge: RSC Publishing.

Farina, Y., Abdullah, M. P., Bibi, N., & Khalik, W. M. A. W. M. (2017). Determination of pesticide residues in leafy vegetables at parts per billion levels by a chemometric study using GC-ECD in Cameron Highlands, Malaysia. *Food Chemistry, 224*, 55−61.

Harvey, D. (2000). *Modern analytical chemistry*. Boston: McGraw-Hill.

International Accreditation Forum. IAF (2020). Publications. <http://www.iaf.nu//articles/Publications/6> Accessed 9.20.

International Conference on Harmonisation of Technical Requirements for Registration of Pharmaceuticals for Human Use. (2005). *Validation of analytical procedures: Text and methodology Q2 (R1)*. Geneva.: ICH.

International Standard Organization. (1993). *ISO 3534-1: Statistics—vocabulary and symbols: Part 1: Probability and general statistical terms*. Geneva.: ISO.

International Standard Organization. (1994). *ISO 5725-3: Accuracy (trueness and precision) of measurement methods and results—part 2: Basic method for the determination of repeatability and reproducibility of a standard measurement method*. Geneva: ISO.

International Standard Organization. (2005). *ISO/IEC 17025: General requirements for the competence of testing and calibration laboratories*. Geneva: ISO.

International Standard Organization. (2016). *ISO Guide 17034: General requirements for competence of reference material producers*. Geneva: ISO.

International Union of Pure and Applied Chemistry. IUPAC. (2020). *IUPAC compendium of chemical terminology—The gold book*. <http://goldbook.iupac.org/index.html> Accessed September.

Martens, H., & Naes, T. (1989). *Multivariate calibration*. Chichester: John Wiley & Sons.

Miller, J. N., & Miller, J. C. (2005). *Statistics and chemometrics for analytical chemistry* (5th ed.). Harlow: Pearson.

Olivieri, A. C., Faber, N. M., Ferré, J., Boqué, R., Kalivas, J. H., & Mark, H. (2006). Uncertainty estimation and figures of merit for multivariate calibration. *Pure and Applied Chemistry, 78*, 633−661.

Organisation for Economic Co-operation and Development. OECD. (1998). *OECD principles on good laboratory practice*. Paris: OECD.

Pierzynski, G. M., Sims, J. T., & Vance, G. F. (2005). *Soil and environmental quality* (3rd ed.). Boca Ranton: CRC Taylor & Francis.

Prichard, E., & Barwick, V. (2007). *Quality assurance in analytical chemistry*. Chichester: Wiley.

Ribani, M., Bottoli, C. B. G., Collins, C. H., & Jardim, I. C. S. F. (2004). Validation in chromatographic and electrophoretic methods. *Química Nova, 5*, 771−780.

Szymánka, E., Gerretzen, J., Engel, J., Geurts, J., Blanchet, L., & Buydens, L. M. C. (2015). Chemometrics and qualitative analysis have a vibrant relationship. *TrAC, Trends in Analytical Chemistry, 69*, 34−51.

US Environmental Protection Agency, EPA (2020). *Quality assurance, quality control and quality assessment measures.* <https://archive.epa.gov/water/archive/web/html/132.html> Accessed 9.20.

Wise, S. A. (2018). What is novel about certified reference materials? *Analytical and Bioanalytical Chemistry, 410,* 2045−2049.

CHAPTER 5
Main analytical techniques

Chemical analyses play an important role in agriculture and related areas, as supporting technologies at all stages of agroindustrial chains as grains, forests, pulp and paper, and agricultural waste, among others sources of agricultural products. Furthermore, chemical analyses give the knowledge of chemical composition and presence or absence of contaminants in food and pollutants in the environment. Then, it ensures the quality and the reliability of agricultural products and processes from the producer to the consumer.

In a general way, performing a chemical analysis, or conducting an analytical process, follows common steps illustrated in Fig. 5.1.

Sampling is the initial step that ensures a sample is representative of the material from which it is taken, paying attention to the need to minimize errors and that the sample is composed of the matrix plus the analyte. The preparation is the stage in which the sample goes through a procedure that aims to make it physically available for separation and/or detection (e.g., grinding, solubilization, digestion, partition extraction), emphasizing that in some techniques, the prepared sample can proceed directly to the detection. In the separation step, the sample is divided into its chemical constituents from solute-solvent interaction mechanisms (e.g., physiosorption and chemisorption), while in the detection step, the

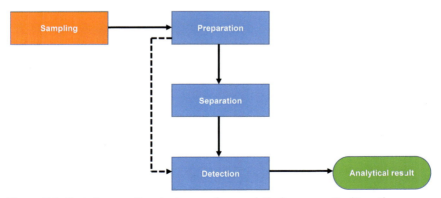

Figure 5.1 Generic operational scheme of an analytical process. *Credit: author.*

intensity of the analyte response to the detector's operational principle (e.g., electric current, absorbed radiation, emitted radiation) which leads to an analytical result to be interpreted to generate the required information—this concept is typical for chromatographic techniques and their methods.

The control and analytical monitoring of agrochemical residues usually requires chemical analyzes that can cover a large number of samples at a low cost (approach 1). On the other hand, more refined studies already require chemical analyzes of complexity and higher costs (approach 2). Thus, these two sets of analytical approaches, which follow pre-established and validated steps, as can be seen in Fig. 5.2. The limits of detection (LOD) and quantification (LOQ) generally have a direct influence on the application of each approach: in the first case (approach 1), the LOQ has greater relevance, while in the second case (approach 2) it is the LOD.

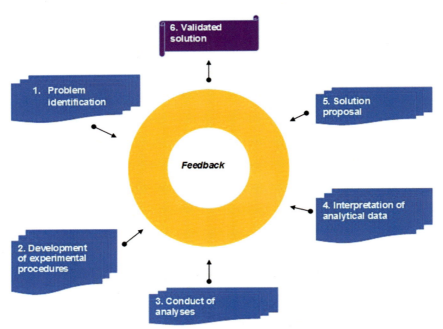

Figure 5.2 Flowchart of the analytical approach to be applied in solving problems arising from agrochemical residues; note that the figure proposes information feedback. *Credit: author.*

There are some considerations to keep in mind about the content of the Fig. 5.2:
- Step 1: problem identification ⇒ an unknown residue in food or in the environment to be analyzed regarding its chemical composition.
- Step 2: development of experimental procedures ⇒ the choice of the most appropriated technique(s) to be applied for the problem (Step 1) and the development and validation of the best method (as seen in the Chapter 4: Fundamentals of analytical chemistry).
- Step 3: conduct of analyses ⇒ carrying on the method from Step 2.
- Step 4: interpretation of analytical data ⇒ what obtained data from Step 3 have to show?—taking into accounting statistics, for example, precision measurements, chemometrics.
- Step 5: solution proposal ⇒ Steps 1−4 showed statistically reliable for the problem solution.
- Step 6: validate solution ⇒ the solution from Step 5 should be evaluated according procedures of validation (seen in the Chapter 4: Fundamentals of analytical chemistry).

As the chapter subject is introduce the main analytical techniques for agrochemical residues in order to carry out analytical studies, we will focus on those most representative classes of analytical instrumentation in order to generate quantitative data whenever possible, as:
- Spectroscopy
- Mass spectrometry
- Chromatography
- Electrochemistry
- Sensors, probes, and bioassays

Furthermore, special attention is dedicated to the extraction techniques.

For a concise understanding of fundamentals of analytical chemistry behind analytical techniques and methods application, some textbooks can be used as:
- Fundamentals of Analytical Chemistry (Douglas A. Skoog, Donald M. West and F. James Holler) (Skoog, West, Crouch, & Holler, 2014).
- Quantitative Chemical Analysis (Harris, 2010).
- However, a good basis is offer in the Chapter 4, Fundamentals of analytical chemistry.

5.1 Spectroscopic techniques

This diverse set of techniques is functionally based on the extent of absorption or emission of electromagnetic radiation by the analyte. When we can correlate the intensity of the signal with the analyte concentration, by means a calibration curve, we have a spectrometric technique.

5.1.1 Absorption of UV–vis radiation, or molecular spectrophotometry

This technique is widely used for the identification and determination of organic, inorganic, and biological species. Usually, molecular absorption spectra are more complex than atomic absorption spectra due to the higher number of energy states of the molecule compared to the isolated atoms (see ahead in the atomic spectrometry item).

The UV region of the electromagnetic spectrum is approximately comprised between 200 and 400 nm and the region of the visible is comprised between 400 and 750 nm. The absorption of radiation by molecules in these regions results from the interactions between photons and electrons that participate in a chemical bond, or between electrons that are not bound in atoms like oxygen, sulphur, nitrogen, and halogens. The wavelength where absorption occurs depends on the type of bond that these electrons participate. Electrons shared in single carbon–carbon or hydrogen–hydrogen bonds are so tightly bound that they require high energy at wavelengths below 180 nm and are not observed by the most common methods of analysis. Due to experimental difficulties in working in this region, single bond spectra are poorly explored. The electrons involved in double and triple bonds are not so strongly trapped and, consequently, they are excited more easily and produce more useful absorption peaks.

Absorption spectroscopy in UV–vis is mainly used in quantitative analysis of several organic compounds containing mainly $C=O$ and $C=C$ bonds, as the intensity of the absorption peaks can be directly correlated to the concentration of the analyte, now called spectrophotometry—is widely used as a detector after separation by liquid chromatography, to be seen later. The signal intensity at a given wavelength value can be directly correlated with the analyte concentration, which allows quantitative data to be obtained—it is worth noting that it is necessary to have the respective curve with linear behavior.

Fig. 5.3 describes the block diagram for a spectrophotometer.

Main analytical techniques 115

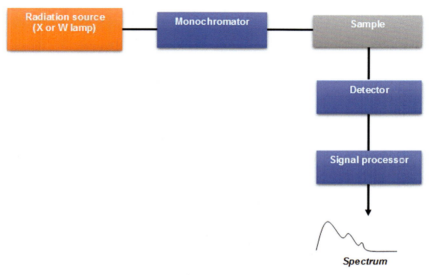

Figure 5.3 Simplified block diagram for a UV–vis spectrophotometer. *Credit: author.*

Table 5.1 Examples of chemical groups present in food, which absorb UV radiation, and their associated electronic transitions.

Chemical group	Structure	Electronic transitions	λ_{max} (nm), nearly
Carbonyl (ketone)	RR'C = O	$\pi \rightarrow \pi^*$	180
		$n \rightarrow \pi^*$	271
Carbonyl (aldehyde)	RHC = O	$\pi \rightarrow \pi^*$	190
		$n \rightarrow \pi^*$	293
Carboxyl	RCOOH	$n \rightarrow \pi^*$	204
Amide	RC = ONH$_2$	$\pi \rightarrow \pi^*$	208
		$n \rightarrow \pi^*$	210
Conjugated diene	RCH-CH = CH-CHR	$\pi \rightarrow \pi^*$	250
Aromatic	C$_6$H$_6$	$\pi \rightarrow \pi^*$	256

Source: Reproduced with permission from, Vaz, Jr. S. (2018). Analytical techniques. In: *Analytical chemistry applied to emerging pollutants*. Cham: Springer Nature.

Table 5.1 describes information on electronic transitions and wavelengths of UV absorption of some chemical groups present in several food samples (Vaz, 2018).

Advantages of UV−vis spectrophotometry are (EAG Laboratories, 2020):
- Fast sample analysis;
- Suitable for a wide variety of analytes;
- User-friendly interface;
- Little maintenance required.

Limitations are:
- Subject to fluctuations from scattered light and temperature changes;
- Relatively low sensitivity;
- Other sample components may cause interferences;
- Not as specific as chromatography, for instance;
- Requires a relatively large sample volume, > 0.2 mL.

As an example of application, Li, Luo, Hu, and Chen (2018) developed a sensitive, rapid, and simple spectrophotometric method based on an Yb^{3+} functionalized gold nanoparticle (AuNPs-Yb) for detection of organophosphorus pesticides (Fig. 5.4).

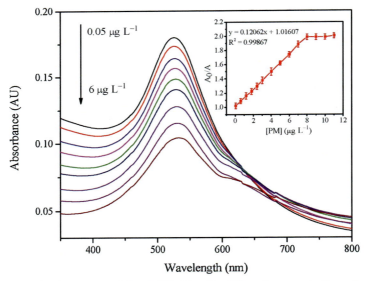

Figure 5.4 UV−vis spectra of the AuNPs-Yb with different concentration of organophosphorus pesticides. Inset: the linear relationship between A_0/A with the concentration of parathion-methyl. *Reproduced with permission from Elsevier, Li, Y., Luo, Q., Hu, R., & Chen, Z. (2018). A sensitive and rapid UV−vis spectrophotometry for organophosphorus pesticides detection based on ytterbium (Yb3 +) functionalized gold nanoparticle. Chinese Chemical Letters, 29, 1845−1848.*

5.1.2 Emission of UV–vis radiation, or fluorescence

The process of UV–vis emission, commonly known as fluorescence, occurs when molecules are excited by absorption of electromagnetic radiation, upon returning to the ground state and releasing the excess of energy as photons.

The sample is excited at a given wavelength, called the excitation wavelength, and its emission is measured at a higher wavelength, called the fluorescence wavelength. This phenomenon is usually associated with systems with electrons π, that is, systems commonly with double bond. Fluorescence usually has a sensitivity and range of work greater than those of UV–vis spectrophotometry. However, it has limited application due to the limited number of systems that fluoresce. It is a technique widely used in the analysis of molecules with aromatic rings, and it is used as a detector after a separation by liquid chromatography. Fig. 5.5 shows a block diagram for a fluorescence spectrometer.

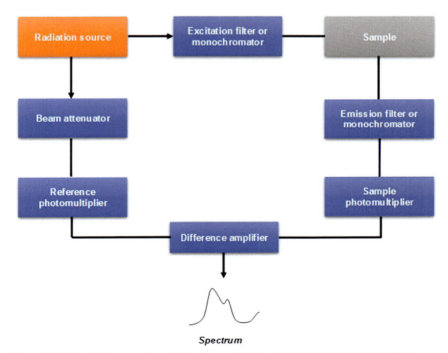

Figure 5.5 Block diagram for a UV–vis spectrofluorophotometer. *Credit: author.*

118   Analysis of Chemical Residues in Agriculture

Pagani and Ibãnez (2019) applied spectrofluorimetric technique in the simultaneous determination of four pesticides in different vegetables and fruits (Fig. 5.6). The method involved non-sophisticated instrumental and simple sample extraction.

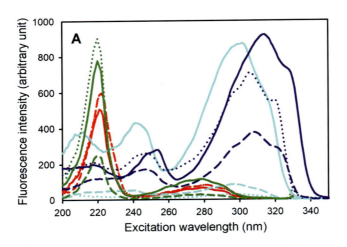

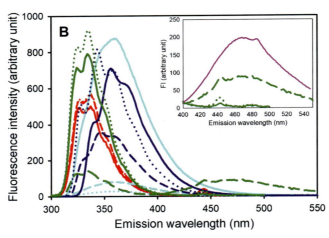

Figure 5.6 Fluorescence spectra of pesticides (A) excitation and (B) emission for TBZ (cyan), NAA (red), FBZ (dark blue) and CBL (dark green); at different pH values: pH = 2 in solid line, pH = 7 in dotted line and pH = 11 in dashed line for all the analytes. Inset emission spectra of CBL (dark green) at different pH values and its metabolite naphthol (pink). TBZ, thiabendazole; NAA, naphthyl acetic acid; FBZ, fuberidazole; CBL, carbaryl. *Reproduced with permission from Elsevier, Pagani, A. P., & Ibãnez, G. A. (2019). Pesticide residues in fruits and vegetables: High-order calibration based on spectrofluorimetric/pH data.* Microchemical Journal, 149, 104042.

5.1.3 Infrared molecular spectroscopy

Vibrational spectroscopy refers to a type of interaction of the radiation with vibrational states of the chemical bonds. Therefore, there is no electronic transition. Here we can highlight infrared (IR) absorption spectroscopy in its three wavelength ranges: near, medium, and far. Polarity has a direct influence on the IR spectrum, modifying its form.

The electromagnetic region of the IR is located between the visible region and the microwaves, that is, from 12,800 to 10 cm^{-1}, remembering that the unit/cm refers to the wavenumber. As discussed above, the IR spectrum is subdivided into three regions: near IR (NIR), medium (MIR, mid-IR) and far (FIR, far-IR). The MIR, which is the most used technique in organic analysis, is divided into two regions: frequency groups, from 4000 up to 1300 cm^{-1}; and absorption of functional groups of two atoms, or vibration, of 1300 to approximately 700 cm^{-1}, also called *fingerprint*. In the NIR the radiation is comprised between 12,800 and 4000 cm^{-1}. The absorption bands in this region are harmonic or combinations of fundamental stretching bands, often associated with hydrogen atoms—this has been a technique in growing in agriculture and related areas, due to the ease of handling of the sample.

IR spectra are typically employed to identify pure organic compounds or impurities, interactions, and binding formation. It is important to consider that a Fourier transform converts the intensity *vs.* time signal into the intensity *vs.* frequency spectrum—from this we have the most used MID technique FTIR (Fourier Transform IR).

Table 5.2 describes the main possible correlations for the assignment of the absorption bands in the MIR as a function of the type of bound. Fig. 5.7 shows the block diagram.

Advantages of FTIR are (EAG Laboratories, 2020):
- Capable of identifying organic functional groups and often specific organic compounds;
- Extensive spectral libraries for compound and mixture identifications;
- Ambient conditions (vacuum is not necessary; applicable for semi-volatile compounds);
- Minimum analysis area: ~15 μm. *Rule-of-thumb: if you can see the sample by eye, it most likely can be analyzed*;
- Can be quantitative with appropriate standards and uniform sample thicknesses.

Table 5.2 Characteristic bands of deformations and vibrational stretches, which may be present in agricultural matrixes.

Band position (cm^{-1})	Assignment	Intensity
3500–3000	Intramolecular stretching of O-H and N-H	Medium absorption
2940–2900	Asymmetric stretching of aliphatic C-H	Strong absorption
1725–1720	Stretching of C = O in COOH and ketones	Strong absorption
1660–1630	Stretching of amide groups (amide band I) and quinone; C = O stretching of hydrogen bonded to conjugated ketones; stretching of COO$^-$	Strong absorption
1620–1600	Stretching of aromatic C = C; stretching of COO$^-$	Medium to weak absorption
1460–1450	Stretching of aromatic C-H	Medium absorption
1400–1390	Deformation of O-H and stretching of C-O and OH phenolic; deformation of C-H in CH$_2$ and CH$_3$; asymmetric stretching of COO$^-$	Medium absorption
1170–950	Stretching of C-O in polysaccharides or polysaccharides-like compounds	Strong absorption

Source: Reproduced with permission from, Vaz, Jr. S. (2018). Analytical techniques. In: *Analytical chemistry applied to emerging pollutants*. Cham: Springer Nature.

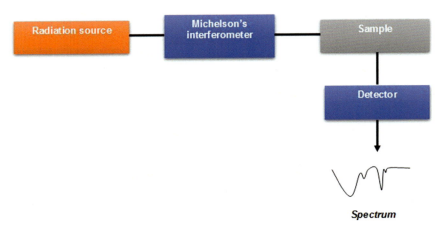

Figure 5.7 Simplified block diagram of a medium infrared spectrometer with Fourier transform (FTIR). Credit: author.

Limitations are:
- Limited surface sensitivity (typical LOD is a film thickness of 25 nm);
- Only specific inorganic species exhibit an FTIR spectrum (for example: silicates, carbonates, nitrates, and sulfates);
- Sample quantitation requires the use of standards;
- Glass absorbs IR light and is not an appropriate substrate for FTIR analysis;
- Water also strongly absorbs IR light and may interfere with the analysis of dissolved, suspended or wet samples;
- Simple cations and anions, for example, Na^+ and Cl^-, do not absorb FTIR light and hence cannot be detected by FTIR; identification of mixtures/multiple sample components may require additional laboratory preparations and analyses;
- Metals reflect light and cannot be analyzed by FTIR.

Fig. 5.7 describe a block diagram.

FTIR can be used to study the presence of residues in the active ingredient (AI) for regulation purposes.

Fig. 5.8 depicts the FTIR spectrum of chlorpyrifos insecticide.

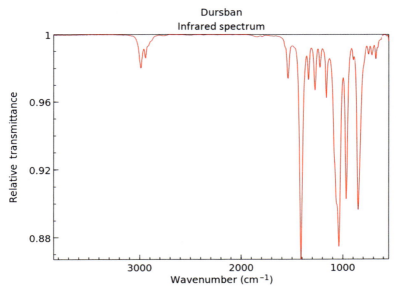

NIST Chemistry WebBook (https://webbook.nist.gov/chemistry)

Figure 5.8 FTIR spectrum for chlorpyrifos molecule obtained in a KBr pellet. *Reproduced with permission from U.S. Secretary of Commerce on behalf of the United States of America.*

5.1.4 Atomic absorption spectrometry

When electromagnetic radiation is applied to atoms in the gaseous state, some of these atoms can be brought to a level of energy that allows the emission of the characteristic radiation of that atom. However, most can remain in the ground state and absorb energy, which in general would correspond to the energy in the gaseous state at the wavelength they would emit if they were excited from the ground state. Thus, when atoms absorb energy, an attenuation of the intensity of the radiation beam occurs. Thus, atomic absorption spectrometry (AAS) is based on the absorption of the electromagnetic radiation by gaseous atoms in the ground state.

AAS is widely used in the inorganic analysis of metals, semimetals, and non-metals in agricultural matrices. There are three different types of atomizer: combustion flame of different gases (hydrogen, acetylene, or natural gas), graphite furnace (or electrothermal) and cold mercury vapor (for determination of the mercury present by reduction to elemental mercury), with the application of each of them depending mainly on the analyte to be determined and the LOD required by the method—the flame AAS is the most common technique. The radiation absorbed has a direct relation with the analyte concentration what turns this technique very useful in quantitative analysis of metals.

In general, the spectra obtained by AAS are simpler than those obtained by atomic or optical emission (see ahead). A particular chemical element absorbs energy at certain wavelengths. Typically, for analysis of an element, the highest absorption wavelength is chosen if there is no interference due to the absorption of the radiation by another element at that wavelength. Due to its simplicity and cost, AAS is the most widely used atomic method. Fig. 5.9 depicts a block diagram for a spectrometer.

Elements frequently detected by AAS are metals and non-metals, except: H, Fr, Ra, Ac, La, Hf, Tc, Os, C, N, O, F, P, S, halogens, and noble gases. As an example of application, Miner et al. (2018) used AAS to determine Fe, Zn, Cu, and Mn concentration (Fig. 5.10) in a long-term study (14 years) about the effects of different nitrogen fertilization—by N-fertilizer supply—on the concentration, uptake, and cycling of N and micronutrients in no-till continuous maize. Moreover, authors assessed nutrient removal via grain harvest.

Main analytical techniques   123

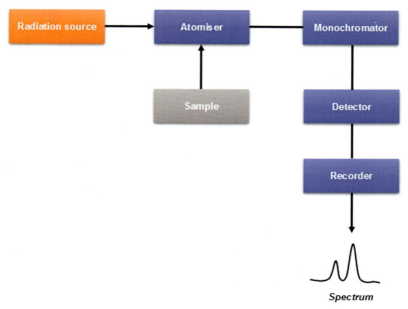

Figure 5.9 Block diagram for an AAS equipment. *Credit: author.*

5.1.5 Atomic emission spectrometry or optical emission spectrometry

Atomic emission spectrometry (AES), or optical emission spectrometry (OES), is based on the measurement of the emission of the electromagnetic radiation in the UV-visible region by neutral and ionized atoms, not in excited state, being widely used in elemental analysis. The most common OES system uses an argon plasma torch that can reach up to 9000K (inductively coupled plasma, ICP) for the electrons excitation in gaseous state. ICP can also be coupled to a quadrupole mass analyzer (ICP-MS); it offers extremely high sensitivity to a wide range of elements.

The technique has high stability, sensitivity, low noise, and low background emission intensity. However, because it involves relatively expensive methods that require extensive operator training, it is not as applied as AAS. All metals or non-metals of agricultural interest, determined by AAS, can be determined by OES—the latter can favor, for some elements, the achievement of lower values of LOD and LOQ.

Advantages of OES are (EAG Laboratories, 2020):
- Bulk chemical analysis technique that can determine simultaneously up to 70 elements in a single sample analysis.

124   Analysis of Chemical Residues in Agriculture

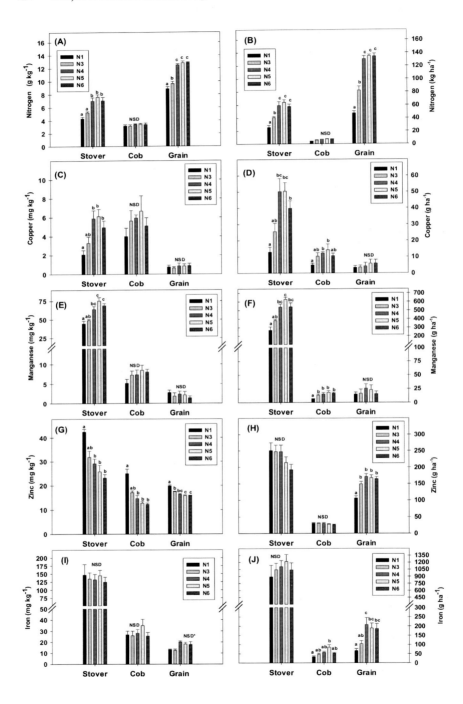

- The linear dynamic range is over several orders of magnitude.
- Instrumentation is suitable to automation, thus enhancing accuracy, precision, and throughput.

Limitations are:
- The emission spectra are complex and inter-element interferences are possible if the wavelength of the element of interest is very close to that of another element.
- In mass spectrometry, determination and quantification of certain elements can be affected by interference from polyatomic species, matrix elements and atmospheric elements.
- The sample to be analyzed must be completely digested, or dissolved prior to analysis in order to determine the element(s) of interest.

5.1.6 X-ray emission spectrometry

This technique allows a rapid and non-destructive multi-element analysis for solid and liquid samples (identification and quantification). When an atom is excited by the removal of an electron from its inner layer, it emits X-rays when returns to its ground state; such radiation has a typical signal intensity for each element, which is used in the analysis.

There are two XRF systems available: the wavelength dispersive spectrometer (WDXRF) and the energy dispersive spectrometer (EDXRF)—the latter has higher signal throughput, which enables small area analysis or mapping.

Advantages of XRF are (EAG Laboratories, 2020):
- Non-destructive technique;
- Can analyze areas as small as $\sim 150 \, \mu m$;

Figure 5.10 Nitrogen (N) concentration (panel A), total N uptake (panel B), and micronutrient concentration (panels C, E, G, I) and uptake (panels D, F, H, J) by plant compartment, averaged over two study years (2013–2014) for five N rates (N1 = 0 kg ha^{-1}, N3 = 67 kg ha^{-1}, N4 = 120 kg ha^{-1}, N5 = 173 kg ha^{-1}, and N6 = 232 kg ha^{-1}). Error bars represent the standard error of the mean (n = 6). Values within plant compartment marked with different letters are significantly different at $\alpha = 0.05$ (Tukey's HSD). Asterisk in panel (I) indicates a significant difference at $\alpha = 0.10$. Grain [Fe] increased with N rate in the mixed model (p = 0.0324), but individual comparisons were significant only at $\alpha = 0.1$ for N1$_{trt}$ versus N4$_{trt}$ (p = 0.0712) and N3$_{trt}$ versus N4$_{trt}$ (p = 0.0538). Note that the y-axis scales vary among nutrients. *Reproduced with permission from Elsevier, Miner, G. L., Delgado, J. A., Ippolito, J. A., Barbarick, K. A., Stewart, C. E., Manter, D. A., ... & D'Adamo, R. E. (2018). Influence of long-term nitrogen fertilization on crop and soil micronutrients in a no-till maize cropping system. Field Crops Research, 228, 170–182.*

- Can analyze any solid material;
- Sampling depth ranging from a few micrometers to several millimeters depending on the material.

Limitations are:
- Cannot detect elements lighter than Al using small spot EDXRF;
- Highest accuracy measurements require reference standards similar in composition and/or thickness to the test sample.

Fig. 5.11 presents a block diagram for an XRF equipment.

Jørgensen, Laursen, Viksna, Pind, and Holm (2005) applied EDXRF in the study of distribution of major and trace elements on a former horticultural soil. The purpose of the study was to combine mapping of soil element concentration levels with multivariate statistics for characterization of soil metal pollution in relation to previous and present land use (Fig. 5.12).

5.1.7 Nuclear magnetic resonance

Unlike other types of spectroscopy, in nuclear magnetic resonance (NMR) it is the nuclei of atoms that absorbs radiation and not their

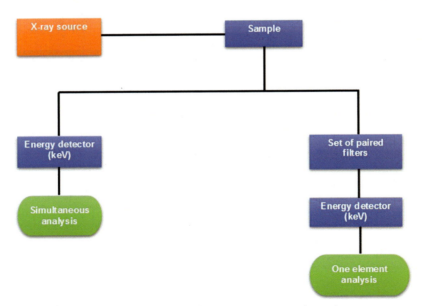

Figure 5.11 Simplified block diagram of an XRF system. *Credit: author.*

Main analytical techniques 127

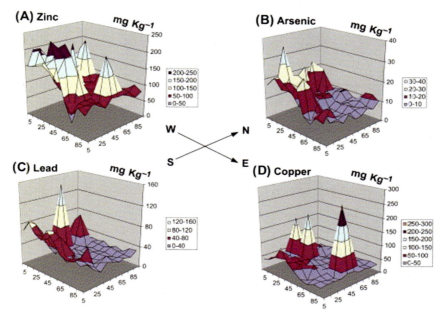

Figure 5.12 Surface distribution of (A) zinc, (B) arsenic, (C) lead, and (D) copper at the investigated area in Vestskoven, Denmark, based on analytical results from EDXRF. *Reproduced with permission from Elsevier, Jørgensen, N., Laursen, J., Viksna, A., Pind, N., & Holm, P. E. (2005). Multi-elemental EDXRF mapping of polluted soil from former horticultural land.* Environment International, 31, 43–52.

electrons. The absorption of radiation by the nuclei occurs when they are subjected to an external magnetic field produced by low energy waves (radiofrequency).

In some cases, the nuclear charge can rotate around the nuclear axis, generating a magnetic dipole. The angular momentum of the moving load can be described in terms of the spin I moment; the most explored nuclei in NMR for agricultural purposes are the ^{1}H and ^{13}C nuclei that have I equal to ½; ^{31}P can be explored in some cases, as in the biomass composition study. The absorption of radio frequency by these nuclei is characteristic and influenced by neighboring nuclei. This allows the molecular structure of a series of chemical compounds to be determined as a function of the chemical displacement (δ) produced according to the electronic density of the atoms present in the molecule.

Fig. 5.13 depicts the block diagram of a NMR spectrometer.

128   Analysis of Chemical Residues in Agriculture

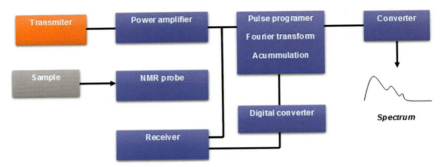

Figure 5.13 Simplified block diagram for a pulsed FT-NMR spectrometer. *Credit: author.*

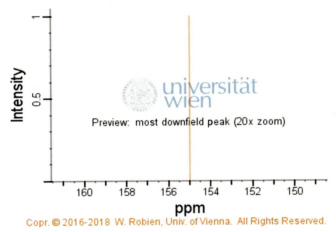

Figure 5.14 ^{13}C-NMR spectrum of carbendazim molecule. *Reproduced with permission from University of Vienna.*

As FTIR, NMR can be used to study the presence of residues in the active ingredient (AI) for regulation purposes. NMR is most largely used to generate qualitative data.

Fig. 5.14 depicts a ^{13}C-NMR spectrum of the carbendazim fungicide.

In a general way, the following assignments of groups as a function of chemical shift can be made for the ^{13}C-NMR spectrum:
- 0−45 ppm: unsubstituted aliphatic C, as in alkanes and fatty acids, due to methyl-terminal groups.
- 45−65 ppm: C associated with N-alkyl, as in amino acids, peptides and proteins and C methoxy.

- 60−110 ppm: C associated with aliphatic O, as in sugars.
- 110−140 ppm: unsubstituted and alkyl substituted aromatic C.
- 110−160 ppm: total aromatic C related to unsubstituted, alkyl substituted and phenolic group.
- 140−160 ppm: C phenolic.
- 160−185 ppm: C in carboxylate.
- 185−230 ppm: ketone C in esters and amides.

5.2 Mass spectrometry

Mass spectrometry (MS) is essentially a technique for detecting molecular components having the mass/charge ratio (m/z) as the unit of measurement. Depending on the ionization technique used, analytes may be present with one or multiple charges. In single charge components, the m/z ratio corresponds to the total mass of the ion in Daltons. In cases where ions with two or more charges are more frequent, the calculation of the original ion mass will depend on deconvolutions of the original signal.

The direct analysis of the sample in the mass spectrometer seldom generates results that can be considered quantitatively, even if the sample is pure. This is a consequence of the high sensitivity of the technique and the efficiency of the ionization process, besides the intrinsic characteristics of each sample that allow greater or less easiness of ionization.

MS is often associated with a separation technique, usually gas chromatography or liquid chromatography—are the *hyphenated techniques*—, where a separation technique coupled to a detection and quantification technique is used. In this case, the mass spectrometer functions as a detector. Such hyphenated techniques make it possible to separate complex mixtures, identify the components, and quantify them in a single operation. Almost all measurements of MS are done under high vacuum, as this allows the conversion of most of the molecules into ions, with a lifetime enough to allow their measurement. The mass spectrometer consists essentially of three components: ionization source, mass analyzer, and ion detector.

There are several commercially available ionization systems: electron impact ionization (EI), chemical ionization (CI), fast atom bombardment (FAB), particle beam bombardment (PBB), matrix assisted laser desorption ionization (MALDI), electrospray ionization (ESI), atmospheric pressure photoionization (API), and atmospheric pressure chemical ionization (APCI). For high molecular weight, nonvolatile and heat sensitive

130   Analysis of Chemical Residues in Agriculture

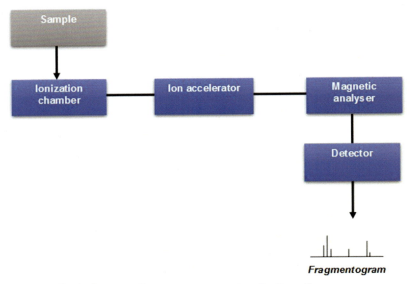

Figure 5.15 Block diagram of a mass spectrometer. *Credit: author.*

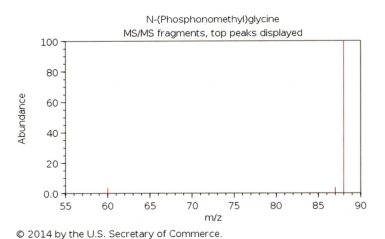

© 2014 by the U.S. Secretary of Commerce.

Figure 5.16 MS/MS fragmentogram of the glyphosate molecule. *Reproduced with permission from U.S. Secretary of Commerce.*

materials, such as some pesticides and food components, MALDI, APCI, and FAB techniques are used. The most common analyzers are: quadrupole, quadrupole ion trap, and time-of-flight tube. The detection is done by electron multiplier tube.

Advantages of high-resolution MS are (EAG Laboratories, 2020):
- Provides comprehensive accurate mass information in a single analysis by MS^n technology (tandem mode);
- Detects more low-level components in complex samples;
- Designed and well suited for large molecule analysis.

Main limitation is that a large volume of data to process requires an experienced analyst to operate.

As FTIR and NMR, MS can be used also to study the presence of residues in the active ingredient (AI) for regulation purposes.

Fig. 5.15 depicts a block diagram.

Fig. 5.16 depicts a fragmentogram for a molecule of the herbicide glyphosate.

5.3 Chromatographic techniques

Chromatography is, conceptually, a technique of separating components from a sample according their *retention time*, for further identification and determination/quantification. It is the most largely used category of instrumentation technique for agrochemical residue analyses due to its versatility.

In the vast majority of cases, chromatographic techniques are coupled with detection techniques, what is known as hyphenated techniques. As forms of hyphenation, we can mention:
- Coupling of solid-phase extraction systems, known as SPE, and SPME (solid-phase microextraction)—these systems allow increased extraction performance from equilibrium phenomena, or thermal sorption-desorption or with organic solvents, which may help to reduce LOD and LOQ values.
- Liquid chromatography (LC) coupling with gas chromatography (GC), or vice versa, promoting the so-called multidimensional separation techniques that allow to work with complex mixtures, such as: LC-GC, LC-GCxGC, LCxLC, etc.; however, the use of chemometrics—seen in the Chapter 4, Fundamentals of analytical chemistry—for the treatment of the generated data is required for this type of hyphenation.

One way to classify the chromatographic techniques is by the physical form of the mobile and stationary phases. Thus, the first classification would be *planar* or *column*—from planar originates the thin layer chromatography and from column liquid and gas phase chromatographies.

Table 5.3 Description of categories of chromatographic techniques according to the stationary phase, considering only the case where the separation takes place in chromatographic columns, which is the type of separation most applied in agricultural matrixes.

General classification	Category	Stationary phase	Equilibrium type
Gas chromatography	Gas-liquid	Liquid bound to solid	Gas-liquid partition
	Gas-solid	Solid	Adsorption
Liquid chromatography	Liquid-liquid partition	Liquid bound or adsorbed to solid	Liquid-liquid partition (immiscible)
	Liquid-solid or adsorption	Solid	Adsorption
	Ion exchange	Resin for ion exchange	Ion exchange
	Size exclusion	Liquid in the interstices of polymeric solid	Partition or penetration
	Affinity	Liquid bound to solid surface	Liquid-liquid partition

Source: Reproduced with permission from, Vaz, Jr. S. (2018). Analytical techniques. In: *Analytical chemistry applied to emerging pollutants*. Cham: Springer Nature.

Table 5.3 provides a description of the functional division categories for GC and LC.

The division presented above is due to physicochemical equilibrium phenomena, which are those that govern the transfer of analyte mass between the mobile and stationary phases. Partitioning is emphasized here, through the chemisorption (involving covalent bonds) and physisorption (involving intra- or intermolecular interactions, usually Van der Walls forces).

5.3.1 Gas chromatography

In GC the components of a sample are separated as a function of their partition between a gaseous mobile phase, usually the helium gas, and a liquid or solid phase contained within the column. One limitation of GC is when the analyte to be analyzed is not volatile (i.e., it is thermally stable); an alternative is the *derivatization*, when the formation of another molecule from the analyte with lower boiling values. The elution of the components is done by an inert mobile phase (carrier gas) flow; that is, the mobile phase does not interact with the molecule of analyte.

The modernization of the equipment, through the development of new stationary phases and data processing software, also led to an investment in systems that provide higher speed during the chromatographic analysis. The shortest analysis time has the direct consequence of reducing the cost of the analytical process and increasing the analytical capacity of the laboratory. The increase in the speed of the chromatographic analysis can be related to the reduction of the size of the column, and reduction of its internal diameter, which compensates the loss of resolution in the determinations.

Regarding the choice of the most suitable detector to be used, the nature of the sample (matrix + analyte) should be taken into account. Several detectors are commercially available for use in GC, with thermal conductivity, flame ionization, electron capture, and mass spectrometer detectors being most commonly used. An ideal detector should meet the following characteristics:

- Adequate sensitivity;
- Good stability and reproducibility;
- Linear response to analytes, extending to several orders of magnitude;
- Temperature range from ambient to at least 400 °C;
- Ease of use;
- Similarity of response to all analytes in the sample.

In practice, the detectors do not group all of the features described above. Table 5.4 shows the most common detectors used in GC and their LOD.

Table 5.4 Most common GC detectors.

Detector	LOD
Flame ionization detector (FID)	0.2 pg
Thermal conductivity detector (TCD)	500 pg
Electron capture detector (ECD)	5 fg
Thermal-ionic detector (TID) or nitrogen-phosphorus detector (NPD)	0.1 pg
Mass spectrometer (MS)	<100 pg

1 pg = 10^{-12} g; 1 fg = 10^{-15} g.
Source: Reproduced with permission from, Vaz, Jr. S. (2018). Analytical techniques. In: *Analytical chemistry applied to emerging pollutants*. Cham: Springer Nature.

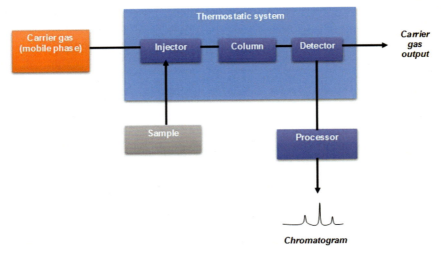

Figure 5.17 Block diagram of a GC equipment. *Credit: author.*

Fig. 5.17 describes a block diagram for a GC system.

Shabeer et al. (2018) developed and reported for the first time a comprehensive multiresidue GC-MS/MS method for 243 pesticides in cardamom—this was associated to QuEChERS extraction (see ahead). The solvent exchange into ethyl acetate reduced the matrix co-extractives and improved sensitivity, observed in the detector response (Fig. 5.18).

5.3.2 Liquid chromatography

LC can be applied in a variety of operating modes, with the best mode depending on the structural characteristics of the analyte to be separated by the chosen analytical method. The most common categories are: partition chromatography—or ion chromatography—, adsorption chromatography, ion exchange chromatography, size exclusion chromatography, and affinity chromatography.

5.3.2.1 High performance liquid chromatography

The use of low-pressure and high-pressure columns, called high performance liquid chromatography (HPLC), outperforms GC in the analysis of semi-volatile and nonvolatile organic compounds. In its many variants, it allows the analysis of complex mixtures, difficult to separate by other techniques, especially mixtures of biomolecules.

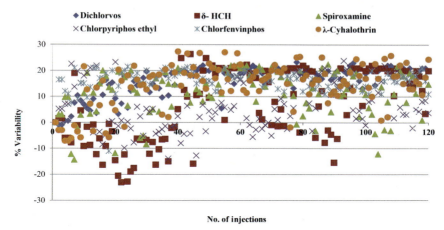

Figure 5.18 Variation in the MS detector response over 120 injections. *Reproduced with permission from Elsevier, Shabeer, T. P. A., Girame, R., Utture, S., Oulkar, D., Banerjee, K., Ajay, D., ... & Menon, K. R. K. (2018). Optimization of multi-residue method for targeted screening and quantitation of 243 pesticide residues in cardamom (Elettaria cardamomum) by gas chromatography tandem mass spectrometry (GC-MS/MS) analysis. Chemosphere, 193, 447−453.*

Typically, the HPLC equipment is equipped with two or more solvent reservoirs. Elution with a single solvent or a mixture of solvents of constant composition is called *isocratic elution*; while the use of a mixture of solvents at different polarity, with composition varying in a programmed manner, is a *gradient elution*. Generally, gradient elution improves the efficiency of the separation process. The pumping system is an important component whose function is to ensure a constant and reproducible flow from the mobile phase to the column. They have a pressure of 0.1−350 bar. The columns are generally stainless steel with lengths ranging from 10 to 30 cm and internal diameters between 2 and 5 mm. The column fillings (or stationary phase) typically have particles with diameters between 3 and 10 μm. Systems with particles smaller than 2 μm and pressures in the range of 1000 bar are called ultra–high-performance liquid chromatography (UHPLC) or ultra-performance liquid chromatography (UPLC). This mode of liquid chromatography can provide a higher resolution in a shorter retention time. Stationary phases for most chromatography modes consist of a silica material, or a polymer such as a polysaccharide or polystyrene, with functional groups of interest attached to the surface of this substrate; they may be either normal phase (polar stationary phase) type or reverse phase (nonpolar stationary phase) type.

Table 5.5 Characteristics of the main HPLC detectors.

Detector	LOD
UV–vis absorption or diode array detector (DAD)	10 pg
Mass spectrometer (MS)	1 pg
Fluorescence detector (FD)	1 ng

1 pg = 10^{-12} g; 1 ng = 10^{-9} g.
Source: Reproduced with permission from, Vaz, Jr. S. (2018). Analytical techniques. In: *Analytical chemistry applied to emerging pollutants*. Cham: Springer Nature.

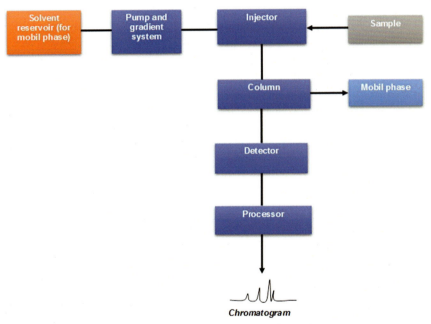

Figure 5.19 Block diagram for a HPLC equipment. *Credit: author.*

Selection of the mobile phase is critical for partitioning, adsorption and ion exchange chromatography, and less critical for the other modes. For the solvents used to form this phase, properties such as the UV–vis cut-off wavelength and the refractive index are important parameters when working with UV–vis and/or refractive index detectors. The polarity index (P') and the eluent force (ε^0) are polarity parameters that aid in choosing the phase for partitioning and adsorption chromatography, respectively.

Main analytical techniques 137

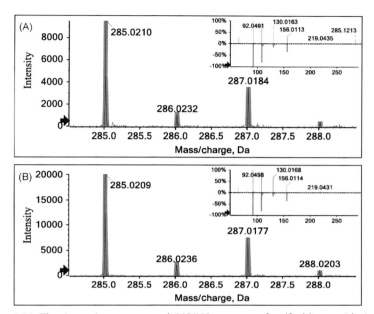

Figure 5.20 The isotopic pattern and MS/MS spectra of sulfachloropyridazine: (A) standard solution; (B) sample. *Reproduced with permission from Elsevier, Hou, X., Xu, X., Xu, X., Han, M., & Qiu, S. (2020). Application of a multiclass screening method for veterinary drugs and pesticides using HPLC-QTOF-MS in egg samples.* Food Chemistry, 309, 125746.

As for GC, there are several types of detectors available commercially, and the choice usually depends on the type of analyte and the number of analyzes required. Detectors may be concentration sensitive, when the analytical signal produced is proportional to the analyte concentration in the effluent or eluted; or mass sensitive, when the signal produced is proportional to the mass flow rate. Table 5.5 lists the main detection systems for HPLC.

Fig. 5.19 depicts a block diagram for a HPLC system.

Hou, Xu, Xu, Han, and Qiu (2020) developed a method based on HPLC-QTOF-MS—HPLC quadrupole-time-of-flight mass spectrometry—for multiresidue analysis in eggs—analytes: 76 veterinary drugs and pesticides residues—, with a single-injection workflow used for screening analysis, with the Fig. 5.20 presenting a mass chromatogram for sulfachloropyridazine, a broad spectrum sulfonamide antibiotic used in veterinary medicine and in the swine and cattle industries.

5.4 Electrochemical techniques

Electrochemistry studies the conversion of electrical energy into chemical energy, and vice versa, considering the transport of charges of ionic species. Some electrochemical techniques are based on oxidation-reduction reactions, such as potentiometry, coulometry, electrogravimetry, and voltametries. Others, in Faradaic processes, as in the case of condutimetry.

5.4.1 Potentiometry

Potentiometry is a technique based on measuring the potential of electrochemical cells without appreciable current consumption. The potentiometric measurements are perhaps the most accomplished in the instrumental chemical analysis, being that of the hydrogenation potential is the best known and applied.

The basic structure of a potentiometer is composed of reference electrode, indicating electrode (or work electrode) and a potential measuring device. Ideally, the reference electrode is a half-cell that has a known and constant electrode potential at a given temperature, independent of the composition of the analyte solution. Potentiometric methods were initially developed to determine the end point of a titration; later, they were used to determine the concentration of ionic species through the so-called direct potentiometry. The technique requires only the comparison of the potential developed in the cell, after immersion of the indicator electrode in the analyte solution, with its potential when immersed in standard solutions of known concentrations of the analyte.

One of the applications of direct potentiometry is the determination of the hydrogen ionic potential (pH) of aqueous media using, for this purpose, a glass electrode and a pH meter. This potentiometric method is possibly the most common analytical method ever created. In infinitely diluted solutions the activity of an ionic species is approximately equal to its concentration. Thus, the concentration of the species to be determined is related to the potential of the electrode.

5.4.2 Voltammetry

Voltammetry is a technique that involves the determination of substances in solution that can be oxidized or reduced on the surface of an electrode. For these determinations the relationships between current, voltage and time during electrolysis in a cell are studied. The equipment for voltammetry employs three electrodes immersed in the solution containing the analyte and an excess of non-reactive electrolyte, called support electrolyte.

The current of analytical interest is the faradaic current, which arises due to the oxidation or reduction of the analyte in the working electrode. The current due to the migration of ions under the influence of an electric field is called the capacitive current. The voltage at the working electrode varies systematically as the current response is measured. Various voltage-time functions called excitation signals can be applied to the working electrode; in function of these signals of excitation is that one has the type of voltammetry: square wave, linear sweep in anodic or cathodic direction, cyclic, and polarography—very useful for inorganic analysis of contaminants or pollutants. The simplest type is linear sweep voltammetry, where the potential in the electrode of work increases or decreases linearly while the current is recorded. With the development of differential pulse and square wave voltammetries it became possible analyte determinations of the order of 10^{-7} to 10^{-8} mol L^{-1}—measurements of lower concentrations are affected by the residual current.

Analytical pre-concentration processes have been used for trace analysis in order to increase the faradaic current. One of the techniques used is the anodic dissolution voltammetry, which can be used in the determination of toxic metals in soil and water.

5.4.3 Electrophoresis

Electrophoresis is an electrochemical separation technique based on the separation of species of different electric charges after the application of an electric field in a conductive liquid known as *electrolyte*. The cations should migrate to the cathode—negatively charged -, while the anions will migrate to the positively charged anode; ions of higher electric charge (e.g., M^{4+}) will migrate faster than those of lower charge (e.g., M^{2+}). Electrophoresis can achieve better resolutions than LC for some chemical species.

In capillary electrophoresis (CE), which is the most widely used mode for environmental analysis, the electrolyte is kept inside a capillary tube of internal diameter between 25 and 75 µm, with the sample being injected into one end of this tube. As the sample migrates through the capillary from the application of the external electric field, its components are separated and eluted at different time values, which results in an *electropherogram*, providing qualitative and quantitative information.

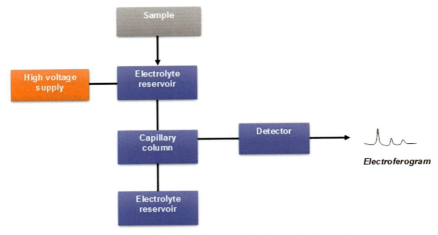

Figure 5.21 Block diagram of a CE equipment. *Credit: author.*

In general, the following observations can be taken into account for the application of electrophoresis:
- Solutes with high electrophoretic mobility will have high separation efficiency;
- Selectivity can often be improved by adjusting the pH values of the conducting medium (electrolyte);
- Resolution is improved by increasing the applied voltage.

Detectors commonly used in capillary electrophoresis are: amperometric, mass spectrometry, fluorescence, and UV–vis absorption. Fig. 5.21 shows the block diagram of the equipment.

Hernández-Mesa, Moreno-González, Lara, and García-Campaña (2019) observed that CE is a well-established separation technique, applied in food analysis because of its advantages related with the high separation efficiency, short analysis time and low cost. Its applications in the field of food analysis cover the determination of relevant contaminants, such as natural toxins or residues of pesticides or veterinary drugs, among others.

5.5 Probes and sensors

In the broadest definition, a sensor is an electronic component, module, or subsystem whose purpose is to detect events or changes in its environment and send the information to other electronics, frequently a computer processor. A sensor is always used with other electronics, whether as simple as a light or as complex as a computer.

A chemical sensor is a self-contained analytical device that can provide information about the chemical composition of its environment, that is, a liquid or a gas phase (mainly) (Banica, 2012). The information is provided in the form of a measurable physical signal that is correlated with the concentration of a certain chemical species (analyte). Two main steps are involved in the functioning of a chemical sensor, namely, *recognition* and *transduction*. In the recognition step, analyte molecules interact selectively with receptor molecules or sites included in the structure of the recognition element of the sensor. Consequently, a characteristic physical parameter varies and this variation is reported by means of an integrated transducer that generates the output signal (the transduction step). A chemical sensor based on recognition material of biological nature is a biosensor. Nowadays, the development of new materials for sensors, as molecularly imprinted polymers and aptamers, has eliminate the differentiation of chemical/biochemical sensors.

Electrochemical sensors are well-recognized as easy to hand and fast devices for environmental and food analyses. For instance, gold nanomaterials based electrochemical sensor systems have a strong potential for environmental monitoring—organic and inorganic pollutants -, trough enhanced and stable analytical capabilities (Jin & Maduraiveeran, 2017). As traditional techniques (e.g., spectroscopies and chromatographies) require either lengthy sample preparation events or complicate instrumentation and hence are time-consuming techniques, electrochemical sensors present as a good alternative, since they are rapid and stable response, with a high sensitivity and selectivity, and ease of miniaturization.

The development of paper-based sensors using microfluidic (*lab on paper*) for environmental analysis has brought the promise of cheap, simple and accessible devices for quick, easy, and in-field detection of pollutants (Meredith et al., 2016). According Meredith et al. (2016), this technique can be applied to metals (e.g., Cu, Cd, Pb, Hg, and Cr), non-metals (e.g., $P_{PO4^{3-}}$, N_{NO3^-}, N_{NO2^-}, N_{NH3}, arsenic and cyanide), phenolic compounds (e.g., phenol, BPA, *m*-cresol, *p*-cresol, catechol, and dopamine) and pesticides (e.g., organophosphate insecticides). For the analyte detection we can use colorimetry (quantitative/semi-quantitative), electrochemical (quantitative), electrochemical conductivity (semi-quantitative/quantitative), chemiluminescence (quantitative) and electrochemiluminescence (quantitative) (Liana, Raguse, Gooding, & Chow, 2012). However, the commercial availability of paper-based sensors remains limited with an expectation of increasing for the next years.

As an example of chemical sensor for pesticide residue, Zhang et al. (2017) synthesized mercaptosuccinic acid capped Mn-doped ZnS quantum dots (ZnS:Mn^{2+} QDs) with bright orange emission quantum—a phosphorescence probe—for thiram fungicide residues identification at fresh fruit peels. The strategy is based on the balance of Ag^{+} induced phosphorescence quenching and the affinity coordinating between Ag^{+} ion and thiram (Fig. 5.22).

Miniaturized probes are frequently based on spectroscopic and electrochemical technologies. For instance, colorimetry can be applied for luminescent probe for field screening of Pb^{2+} in water, with a simple, fast, cost-effective and highly selective, without requirements for additional instrumentation (Zeng et al., 2017).

A practical example of chromatographic technique replacement by probe for analysis is the use of photochemically induced fluorescence matrix data combined with second-order chemometric analysis for the

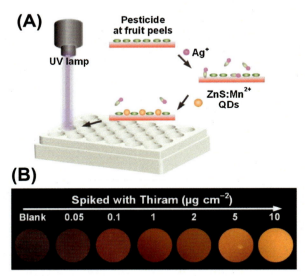

Figure 5.22 (A) Schematic drawing for the on-site detection of thiram at fruit peels. (B) Digital photos of the apple peels in the presence of various amounts of thiram under 312 nm UV light illumination. *Reproduced with permission from Elsevier, Zhang, C., Khang, K., Zhao, T., Liu, B., Wang, Z., & Zhang, Z. (2017). Selective phosphorescence sensing of pesticide based on the inhibition of silver(I) quenched ZnS:Mn2+ quantum dots.* Sensors and Actuators B: Chemical, 252, 1083−1088.

determination of carbamazepine, ofloxacin and piroxicam in water samples of different complexity without the need of chromatographic separation. It was verified that the proposed strategy was simpler and greener than LC-MS methodologies, without compromising the quality of results (Hurtado-Sánchez et al., 2015).

On the other hand, the use of probes for high-throughput analysis of non-target and suspect analytes is very useful because it can minimize costs and time related to the analytical process, besides decrease the response time to combat negative effects on the environment and public health from pollutants.

5.6 Bioassays

Bioassays could provide an alternative to conventional analytical techniques, according green chemistry principles (seen in the Chapter 4: Fundamentals of analytical chemistry), with a minimization of the negative environmental impacts from analytical chemistry. Furthermore, this kind of analysis provides information about the effect of agrochemical residues on the organism and ecosystems what are strongly relevant for the toxicity determination and understanding. Commonly, bioassays are treated as ecotoxycological assessment.

Wieczerzak, Namiésnik, and Budlak (2016) observed this opportunity by means the application of commercially available certified bioassays using single-celled organisms (e.g., cell lines, yeast, bacteria) and multicellular organisms (e.g., invertebrate and vertebrate animals and plants). The main advantages for these bioassays are:
- It is possible to conduct tests in situ;
- It is not necessary to purchase high-purity reagents and reference materials;
- Relatively low cost per analysis.
  As main disadvantages:
- Experience required to select battery of bioassays;
- More difficult to maintain reproducibility and repeatability;
- Necessity to keep clean cultures up.

Besides, the use of microorganisms and organisms for bioassays are object of regulatory surveillance.

Table 5.6 presents an overview of bioassays available for analysis for ecotoxycological purposes.

Table 5.6 Some bioassays commercially available for ecotoxycological assessment.

Bioassay —company	Species used	Measure of toxicity effects	Application	In accordance with
Plant-based bioassays				
Lemna Test—LemnaTec GmbH, Germany	*Lemna minor, Lemna gibba* (duckweed)	Acute and sub-chronic by means growth inhibition in 7 days	Pesticides and other chemicals soluble in water (aquatic samples)	ASTM, US-EPA, OECD
AlgalToxiit F™—MicroBioTests Inc., Belgium	*Raphidocelis subcapitata* (microalgae)	Short-chronic, by means inhibition of growth in 72 h	Pure compounds, effluents, sediments, surface and groundwater, wastewaters	ISO, OECD
AlgalToxiit F™—MicroBioTests Inc., Belgium	*Phaeodactylum tricornutum* (diatom)	Short-chronic, by means inhibition of growth in 72 h	Chemicals released in aquatic as well as terrestrial environments	ISO, OECD
Animal-based bioassays				
DaphtoxKit F™ magna—MicroBioTests Inc., Belgium	*Daphnia magna* (crustaceans)	Acute and chronic toxicity screening test in 24–48 h by means immobilization or mortality, inhibition of reproduction, inhibition of reproduction, inhibition of growth population	Chemicals released in aquatic as well as terrestrial environments	OECD, ISO, US-EPA, ASTM

Rotoxkit F™ short-chronic—MicroBioTests Inc., Belgium	*Brachinious calyciflorus* (cotifers)	Acute and chronic toxicity in 24–48 h by means mortality or reduction of reproduction	Testing toxicity of all chemicals and wastes released in aquatic as well as terrestrial environments	OECD, ISO, US-EPA
DaphtoxKit F™ pulex—MicroBioTests Inc., Belgium	*Daphnia pulex* (crustaceans)	Acute toxicity in 48 h by means immobilization or mortality, with calculation of the EC50 or LC50	Chemicals in wastewater, surface and deep sea waters	OECD, ISO

Single-cell organisms-based bioassays

Microtox® M500—Modern Water, UK	*Photobacterium phosphoreum*	Acute toxicity in 5–30 min by means the decrease in the bioluminescence	Monitoring toxicity of chemicals in water, sediments, soil, etc.	ISO
UmuC Easy AQ—Xenometrix AG, Switzerland	Mutant strains of *Salmonella typhimurium* TA1535/pSK10002	Mutagenicity/genotoxicity in 30 h	Aqueous and concentrated samples of pure compound, wastewater, drinking water	ISO

ASTM, American Society for Testing and Materials; *ISO*, International Standard Organization; *OECD*, Organization for Economic Co-operation Development; *U.S. EPA*, United States Environmental Protection Agency.
Source: Adapted from Wieczerzak, M., Namieśnik, J., & Budlak, B. (2016). Bioassays as one of the green chemistry tools for assessing environmental quality: A review. *Environmental International*, 94, 341–361.

5.7 Sampling

These sampling techniques are divided into two classes: sampling for environmental samples and sampling for food samples.

5.7.1 Environmental samples

Soil samples are collected using the following procedure (U.S. Environmental Protection Agency, 1997):

1. Carefully remove the top layer of soil to the desired sample depth with a precleaned spade.
2. Using a precleaned, stainless-steel scoop, spoon, trowel, or plastic spoon, remove and discard the thin layer of soil from the area that came into contact with the shovel.
3. Transfer the sample into an appropriate container using a stainless-steel or plastic lab spoon or equivalent. If composite samples are to be collected, place the soil sample in a stainless-steel or plastic bucket and mix thoroughly to obtain a homogeneous sample representative of the entire sampling interval. Place the soil samples into labeled containers. [Caution: never composite volatile organic analysis (VOA) samples].
4. VOA samples should be collected directly from the bottom of the hole before mixing the sample to minimize volatilization of contaminants.
5. Check to ensure that the VOA vial Teflon liner is present in the cap, if required. Fill the VOA vial fully to the top to reduce headspace. Secure the cap tightly. The chemical preservation of solids is generally not recommended. Refrigeration is usually the best approach, supplemented by a minimal holding time.
6. Ensure that a sufficient sample size has been collected for the desired analysis, as specified in the sampling plan.[1]
7. Decontaminate equipment between samples.
8. Fill in the hole and replace grass turf, if necessary.

   After the sampling, samples should obey the follow requirements:

- Holding time: 14 days from sampled
- Minimum volume: 100 g

---

[1] A plan containing, basically, amount of samples, location, methods, QA/QC, and logistic of sampling. Steps related to the sample plan are: (1) identify the parameters to be measured, the range of possible values, and the required resolution; (2) design a sampling scheme that details how and when samples will be taken; (3) select sample sizes; (4) design data storage formats; (5) assign roles and responsibilities.

- Container type: two 40-mL vials and no air space (for volatile compounds); glass jar with Teflon-lined cap (for other compounds)
- Preservation: cool to 4°C (ice in cooler)

The sample preparation for the analysis step comprises:
1. Extraction by means solvent (solid—liquid, Soxhlet, accelerated solvent extraction, etc.), according the analyte and matrix physicochemical properties.
2. Clean-up by means the use of solid-phase extraction (SPE) or another technique.
3. Injection for the chromatographic analysis, when it is indicated.

The sampling of surface water takes into account (U.S. Environmental Protection Agency, 2016):
1. Surface water samples will typically be collected either by directly filling the container from the surface water body being sampled or by decanting the water from a collection device such as a stainless steel scoop or other device.
2. During sample collection, if transferring the sample from a collection device, make sure that the device does not come in contact with the sample containers.
3. Place the sample into appropriate, labeled containers. Samples collected for volatile organic compounds (VOCs) analysis must not have any headspace. All other sample containers must be filled with an allowance for ullage.
4. All samples requiring preservation must be preserved as soon as practically possible, ideally immediately at the time of sample collection. If preserved VOCs vials are used, these will be preserved with concentrated hydrochloric acid prior to departure for the field investigation. For all other chemical preservatives, will use the appropriate chemical preservative generally stored in an individual single-use vial. The adequacy of sample preservation will be checked after the addition of the preservative for all samples, except for the samples collected for VOCs analysis. If it is determined that a sample is not adequately preserved, additional preservative should be added to achieve adequate preservation.
5. All samples preserved using a pH adjustment (except VOCs) must be checked, using pH strips, to ensure that they were adequately preserved. This is done by pouring a small volume of sample over the strip. Do not place the strip in the sample. Samples requiring reduced temperature storage should be placed on ice immediately.

These recommendations can be extended to wastewater and sewage.

For groundwater, sampling recommendation comprises (U.S. Environmental Protection Agency, 2013):

1. Groundwater samples will typically be collected from the discharge line of a pump or from a bailer, either from the pour stream of an upturned bailer or from the stream from a bottom-emptying device. Efforts should be made to reduce the flow from either the pump discharge line or the bailer during sample collection to minimize sample agitation.
2. During sample collection, make sure that the pump discharge line or the bailer does not contact the sample container.
3. Place the sample into appropriate, labeled containers. Samples collected for VOC, acidity and alkalinity analysis must not have any headspace. All other sample containers must be filled with an allowance for ullage.
4. All samples requiring preservation must be preserved as soon as practically possible, ideally immediately at the time of sample collection. If preserved VOC vials are used, these will be preserved with concentrated hydrochloric acid prior to departure for the field investigation. For all other chemical preservatives, will use the appropriate chemical preservative generally stored in an individual single-use vial. The adequacy of sample preservation will be checked after the addition of the preservative for all samples except for the samples collected for VOC analysis. If additional preservative is needed, it should be added to achieve adequate preservation.

Furthermore, according the U.S. special sample handling procedures should be instituted when trace pollutants samples are being collected. All sampling equipment, including pumps, bailers, water level measurement equipment, etc., which comes into contact with the water in the well must be cleaned in accordance with cleaning procedures. Pumps should not be used for sampling unless the interior and exterior portions of the pump and the discharge hoses are thoroughly cleaned. Blank samples should be collected to determine the adequacy of cleaning prior to collection of any sample using a pump other than a peristaltic pump.

For organic residues, the best preservation method is cool to 4°C, add sulfuric acid to pH value below 2, using a glass container. The holding time is 28 days. In the case of metals (inorganic residues), generally samples will be filter on site, add nitric acid to pH value below 2, using a glass or plastic container. The holding time is 6 months.

After the sampling, samples will pass by the follow steps:
1. Analyte extraction, using liquid-liquid extraction with a solvent, solid-phase extraction (SPE) or solid-phase microextraction (SPME).
2. Clean-up before the extract injection in the chromatographic system, by means a SPE—when it is indicated.

5.7.2 Food samples

According Dehotay and Cook (2015), analysts should ensure that they are fully aware of parameters of sample heterogeneity with the goal to avoid incorrect sampling and comminution and reduce the uncertainty introduced to test samples. Validation of sampling and sample processing protocols and ongoing QC should be implemented in monitoring programs, just as such data quality assurances are required for analytical methods. If the state-of-the-art two-step comminution process using cryomilling can be demonstrated to yield acceptably accurate results for given analytes/matrices of interest, then the automated high-throughput methods involving microsubsampling may be used just as any other validated protocols. In any case, the dearth of publications in the scientific literature on the subject does not instill confidence among residue chemists and regulators that the described practices are ready for wider implementation. Just as the agrochemical industry led the way when LC-MS/MS was first introduced to help convince others of its attributes and validity, the analytical chemists conducting field trials for pesticide registrations are the early adopters to assess the new cryomilling and automated high-throughput approaches.

For ongoing evaluations, again Dehotay and Cook (2015) suggested that laboratories add QC spikes to samples prior to each step in the analytical process comprised by:
- Comminution
- Cryomilling
- Sample preparation
- And analysis

Thereby isolate these steps to assess the degree of uncertainties and help with troubleshooting if problems occur. The expenditure of so much time, effort, and expense in residue monitoring programs warrant assurances that the analytical results are meaningful for the intended purposes of regulatory registrations, enforcement, risk assessment, and other needs.

For more information about food sampling, the following document from Food and Agriculture Organization of United Nations is recommended:
- Recommended Methods of Sampling for the Determination of Pesticides Residues for Compliance with MRLs CAC/GL 33−1999.[2]

5.8 Sample preparation

Extraction, distribution or partition is the apportionment of a solute—the analyte—between two phases. It take into account equilibrium phenomena summarized in the partition ratio, K_D Eq. (5.1). The ratio of the concentration of a substance in a single definite form, A, in the extract to its concentration in the same form in the other phase at equilibrium, for example, for an aqueous/organic system (International Union of Pure & Applied Chemistry, 2020) is defined as:

$$K_D^A = \frac{[A]_{\text{org}}}{[A]_{\text{aq}}} \tag{5.1}$$

K_D is sometimes called the distribution constant. The use of the inverse ratio (aqueous/organic) may be appropriate in certain cases, for example, where the organic phase forms the feed but its use in such cases should be clearly specified. The ratio of the concentration in the denser phase to the less dense phase is not recommended as it can be ambiguous.

Extraction is a paramount step of the analytical process, especially in the sample preparation, and the use of solvent extraction is the more recurrent in the analysis of organic molecules in several matrixes. According to the International Union of Pure and Applied Chemistry (2020), the solvent extraction can be defined as the process of transferring a substance from any matrix (e.g., water or soil) to an appropriate liquid phase (e.g., a mobile phase for HPLC). If the substance is initially present as a solute in an immiscible liquid phase the process is synonymous with liquid–liquid extraction. Notes: if the extractable material is present in a solid (such as a soil) the term leaching may be more appropriate. The extractable material may also be a liquid entrapped within or adsorbed on a solid phase.

After the extraction step, we can concentrate the analyte present in the extraction medium to promote a better analytical response. It can be made by means (Mitra, 2003):

[2] http://www.fao.org/input/download/standards/361/CXG_033e.pdf

- Stream of nitrogen gas flow, for nonvolatile analyte and small volume to reduce.
- Rotary vacuum evaporator, for large volume to reduce.
- Kuderna-Danish concentrator using air-cooled condenser, for smaller volume to be reduced to less 1 mL.

After the extractive concentration, a *clean-up* step is desirable to remove interfering species previously the chromatographic or electrophoretic separation. These interfering species are very common in food and environmental analysis, mainly for soil samples due to the matrix heterogenic composition. To overcome these difficulties we can use:

- Gel-permeation chromatography (GPC), for the elimination of lipids, proteins, polymers, copolymers, natural resins, cellular components, viruses, steroids, and dispersed high-molecular-weight compounds from the sample. This method is appropriate for both polar and nonpolar analytes.
- Solid-phase extraction cartridges (SPE), for steroids, esters, ketones, glycerides, alkaloids, and carbohydrates. Cations, anions, metals, and inorganic compounds are also candidates for this technique.

As an example of preparative strategies for agrochemical residue analysis—organophopsphorous residue in food by means the application of colorimetry—, Chawla, Kawshik, Swaraj, and Kumar (2018) described the following isolation and fractionation techniques:

- Supercritical fluid extraction (SFE)
- Solid phase extraction (SPE)
- Solid phase micro-extraction (SPME)
- Microwave assisted extraction (MAE)
- Dispersive liquid-liquid micro-extraction (DLLME)
- Matrix solid phase dispersion (MSPD)
- Accelerated solvent extraction (ASE)
- QuEchERS (Quick, Easy, Cheap, Effective, Rugged and Safe)

The choice of the best technology depends on physicochemical properties of analyte and matrix. Aspects of costs and supply availability should be evaluated also. Furthermore, the recovery of the extraction method must be in the range of 70%−120% (seen in the Chapter 4: Fundamentals of analytical chemistry).

Table 5.7 describes extraction techniques for pesticides residues in aqueous environment. These techniques include solid-phase extraction (SPE), dispersive solid-phase extraction (dSPE), stir bar sorptive extraction (SBSE), magnetic solid-phase extraction (MSPE), solid-phase microextraction

Table 5.7 Extraction techniques for pesticide residue analysis in aqueous matrices.

Extraction technique	Adsorbent/extraction solvent	Matrix	Analyte	Analytical technology	LOD (μg L^{-1})	LOQ (μg L^{-1})	LR (μg L^{-1})	R (%)
SPE	MCM	Environmental water	OPPs, carbamates and triazoles	HPLC-MS	0.01	0.05	0.10–75.55	65–126
	Mesoporous silica-based on the sol-gel material CNPrTEOS	Environmental water	Polar and nonpolar OPPs	HPLC-UV	0.072–0.091	0.24–0.30	0.8–100	80.1–108
	HPGA	Environmental water	Pyrethroids	GC-MS	0.012–0.11	—	0.2–50	65.7–105.9
	MIP	Environmental water	Triazines	HPLC-MS/MS	0.007–0.068	0.022–0.227	0.25–50	88–100
	MWCNTs	Environmental water	Carbamate insecticides	HPLC-MS	0.01–0.05	0.08–0.2	5–1000	92.2–103.9
	MIP	Environmental water	Chloroacetamide herbicides	HPLC-DAD	0.03–0.06	0.09–0.15	0.1–200	82.1–102.9
	MIP	Environmental water	Triazines	HPLC-UV	0.05–0.1	—	0.1–25	—
	HLB	Field water	Neonicotinoids, OP and OC insecticides	GC-MS	0.0021–0.0042	—	—	63–124
	COF	Environmental water, juice, fruit and vegetables	Pesticide	HPLC-UV	0.02–0.050	0.06–0.150	0.1–160.0	87–103.3
	ZMNIC	Water and juice	OPPs	GC-MS	0.001–0.004	0.003–0.012	0.01–500	91.9–104.4
	MIP	Maize, water, and soil	Triazine herbicides	HPLC-UV	0.0009–0.0012	—	0.1–16 μmol L^{-1}	82–99
		River 4water	Imidazole fungicides	HPLC-UV	0.023–0.031	—	0.1–20	84.2–95.0

	Waste water and plant	OP insecticides	GC-FPD	294	980	0.9992	94–104
Oasis HLB	Water and food	Carbendazim	HPLC-DAD	7	23	50–5000	84.32–99.14
	Environmental water	Atrazine	HPLC-DAD	0.20	0.60	–	83–89
	Water	Pesticides	HPLC-MS/MS and GC-MS/MS	–	0.01–0.1	LC-MS/MS: 3.5–600 GC-MS/MS: 5–50	LC-MS/MS: 51.9–144 GC-MS/MS: 45–134
C18, PS2, Oasis HLB, AC2, and PLS–3	Water	Pesticides	HPLC-MS	0.008–0.08	–	–	More than 50%
G-coated cotton fiber	Water	Pesticides	GC-MS	–	0.02–0.09	0.02–10	83–107
Oasis HLB	Water	Pesticides	GC-MS/MS and HPLC-MS/MS	0.006	0.02	0.99	70–117.3
CTAB modified zeolite NaY	Water	Pesticides	HPLC-UV	0.005–140	0.02–600	0.9963–0.9991	77–111
Surfactant	Water	Pesticides	HPLC-MS/MS	–	0.027–0.090	0.1–40	83–93
G- modified TiO$_2$	Water	Carbamate pesticide	HPLC-UV	2.27–3.26	–	5–150	83.9–108.8
MOF	Water	OCPs	GC-MS	0.0025–0.016	0.0010–0.074	0.05–50	87.6–98.6
MWCNTs	Surface water	Pesticides	HPLC-MS/MS	0.0003–0.00095	0.0016–0.0452	0.99	100
COOH-PAN NFsM	Surface water	Atrazine	HPLC-DAD	0.09–0.12	0.3	0.3–40	81.35–120.32
HP-β-CD	River water and vegetable juices	Bio pesticides	HPLC-MS	–	3.73–16.51	5–500	88.6–95.8
PLRP-s cartridge	Drinking, ground and surface water	Acidic and polar pesticides	HPLC-MS/MS	0.0012–0.018	Below 0.1	0.05–2	70–120
β-CD polymer @Fe$_3$O$_4$	Honey, tomato, and environmental water	Benzoylurea insectirides	HPLC-DAD	0.02–0.05	–	0.1–80.0	87.3–112.5

(*Continued*)

Table 5.7 (Continued)

Extraction technique	Adsorbent/ extraction solvent	Matrix	Analyte	Analytical technology	LOD (µg L⁻¹)	LOQ (µg L⁻¹)	LR (µg L⁻¹)	R (%)
	C18	Surface waters	OCPs	GC-MS	–	0.00001–0.00037	–	82–117
		Environmental water	Herbicides	HPLC-MS/MS	–	–	–	100
		Surface water	OCPs	GC-MS/MS	0.00008–0.00038	–	0.9981–0.9998	70.3–115.1
		River water	Pesticide	HPLC-UV	0.0028–0.0073	0.0095–0.0244	10–200	88–102
		Surface water	OCPs	GC-ECD	0.0003–0.0054	–	0.99	95–106
	GO	Water and celery	Phenylurea herbicides	HPLC-DAD	0.01–0.02	0.03–0.08	0.05–40	86–112
	Oasis HLB	Drinking water	Neonicotinoid pesticides	HPLC-MS/MS	0.001–0.004	1–200	–	–
		Seawater	Triazine herbicides	HPLC-MS/MS	0.008–0.02	0.023–0.657	0.025–0.5	80.3–99.8
		Surface and ground-water	Pesticides	HPLC-MS/MS	0.001–0.005	0.003–0.01	–	Higher than 70
		Urban wastewater	Pesticides	HPLC-MS/MS	–	0.001–0.015	0.0011–1.6	75–115
		Surface water	Pesticides	GC × GC-MS	1.48	–	–	–
			Pesticides	HPLC-MS	–	0.0005–0.1	0.1–1000	35–89
		Environmental water	Neonicotinoids	HPLC-MS/MS	0.01119	0.00005–0.0005	0.00005–0.1	58.9–109.9
	Seawater	Pesticides	HPLC-MS/MS	0.00002–0.00371	0.00005–0.027	0.0000025–0.4	80–120	83–106
	Poly-Sery HLB	Surface water and sediment	Neonicotinoids insecticides	HPLC-MS/MS	–	0.00001–0.00005	0.99	
	MOF	Cabbage and water	Carbamate pesticides	HPLC-UV	0.01–0.02	–	0.05–20	94.4–99.6
	MOF	Mixed juice, orange juice, and tap water	Phenoxyacetic acid herbicides	HPLC-UV	0.1–0.5	0.2–2	0.2–250	77.1–109.3
	SPE cartridge	Drinking water	Pesticide	GC-MS HPLC-MS/MS	0.00105–0.04821	0.00319–0.14624	0.001–0.25	55.54–121.21

Oasis HLB Oasis WAX Sep-Pak Plus AC 2	Surface water	Pesticide	HPLC-MS/MS	0.00008–0.04433	—	0.2–200	50–150
—	Whole water	Pesticide	HPLC-MS/MS	0.0001–0.1	0.0002–0.005	0.998	63–113
Mesoporous silica doped with Ti	Environmental water	OPPs	GC-NPD	0.4–3.1	0.5–4.4	—	81–104.5
Magnetic hollow zein nanoparticles	Water and soil	Chlorpyrifos	HPLC-UV	25	—	50000–2000000	94.3–96.2
PCP-Cs	Water and vegetable	Phenylurea herbicides	HPLC-UV	0.01–0.02	0.03–0.06	0.1–80.0	97.5–112.8
GO@PA	Environmental Water and leaf lettuces	Phenylurea herbicides and phenylurea insecticides	HPLC-UV	0.10–0.25	0.3–0.75	2.0–100	85.5–106.0
GO-Chm	Rice and river Kabul water	Phenylurea herbicides	HPLC-UV	2.5–3.6	8.6–19	50–6000	90.32–94.09
GO/HIPEs	Farmland water	Triazine herbicides	HPLC-DAD	2.5–5.6	—	25.0–500.0	—
NH$_2$@COF	Environmental water	Carboxylic acid pesticides		0.010–0.060	0.04–0.20	0.2–100	89.6–102.4
HLB	Water	Pesticides	HPLC-MS	0.000001–0.04127	0.000004–0.16508	0.9906	76–97
SDB-RPS and C18	Surface water	Pesticides	GC-MS HPLC-MS	0.03–0.172	0.017–0.099	—	75.33–104.99
Mesoporous silica material	Environmental waters	Pesticides	HPLC-MS/MS	—	0.01	1–100	70.1–113.5
MWCNT	River water	OPPs	HPLC-MS	0.00002–0.001	0.00007–0.333	0.001–0.2	88.3–124.9
PMO	Wastewater	Phenoxy acid herbicides	CE	700–1500	2200–5000	2000–30000	78.3–107.5

(*Continued*)

Table 5.7 (Continued)

Extraction technique	Adsorbent/extraction solvent	Matrix	Analyte	Analytical technology	LOD (μg L^{-1})	LOQ (μg L^{-1})	LR (μg L^{-1})	R (%)
	HBPE	Environmental water	Benzoylurea insecticides	HPLC-UV	0.024–0.068	0.082–0.230	1–25000	85.5–96.12
	MMT	Fruit juice and natural surface water	Neonicotinoid Insecticide	HPLC-DAD	0.005–0.065	0.008–0.263	0.5–1000	8–176
	PPPDA/PVA	Drinking water and lemon and orange juice	OPPs	HPLC-DAD	0.15	0.5	0.5–500	76–102
dSPE	Glass fiber β-CD/ATP	Water Environmental water	Pesticide Pyrethroids	HPLC-MS/MS HPLC-UV	0.0001–0.01 0.15–1.03	0.0005–0.025 0.5–3.43	0.999 2.5–500	77–117 76.8–86.5
	ATP/Fe$_3$O$_4$/PANI	Environmental water	Benzoylurea insecticides	HPLC-DAD	0.02–0.43	—	0.9985–0.9997	77.37–103.69
	M-MWCNT	Environmental water	OCPs	GC-μECD	0.00004–0.00027	0.0001–0.089	0.0002–2	83–115
	G	Environmental water	Triazine and Neonicotine Pesticides	GC-MS and HPLC-MS/MS	0.03–0.4	0.10–1.3	0.5–100	83.0–108.9
	RGO-ZnO	Apple, cucumber and water	OPPs	GC-MS	0.00001–0.00005	0.00005–0.00018	0.0005–0.2	75–104
	N-octylated magnetic NPs	Drinking and surface water	OPPs	GC-FPD	0.02–0.1	0.05–0.33	0.5–800	82–110
	Zinc-based MOF	Water and fruits juice	OPPs	GC-FID	0.03–0.21	—	0.1–100	91.9–99.5
	MOF	Water	Neonicotinoid insecticides	GC-MS	0.02–0.4	0.05–1.0	10–500	73.7–119.0
	Ch-Si MNP	Water	Pesticides	HPLC-VWD	2–46	6–154	12.5–150	100
	GO@ IL@ NHC–Cu	Environmental water	Triazole fungicides	IMS	0.18	0.6	0.994	92–94
	LMA-HEDA	Environmental water	Phenylurea herbicides	HPLC-UV	0.027–0.053	—	0.1–4.0	80.1–97.9

MSPE	PP-CMP	Environmental water	Herbicides	HPLC-MS	0.00055–0.0384	0.00183–0.0126	0.05–10	86.9–101.3
	AF-MCF	Environmental water	Atrazine	HPLC-UV	0.5	—	2–200	94.6–107.2
	mMFP	Tap and river water	Chlorinated herbicides	HPLC-UV	0.04–0.18	0.13–0.60	2–200	87.4–109.5
	IL-mG	Surface water	Triazine herbicides	HPLC-UV	0.09–0.15	0.31–0.51	0.55–500	97–100.8
	Fe$_3$C/MnO/GC	Water, grape, and potato	Herbicides	HPLC-UV	0.01–0.10	0.00004–0.00033	0.1–200	81–112
	MWCNTs	Environmental water	Sulfonylurea herbicides	HPLC-DAD	0.01–0.04	0.03–0.13	0.05–5.0	76.7–106.9
	MOF	Environmental water	Pyrazole/pyrrole pesticides	HPLC-DAD	0.3–1.5	1.0–5.0	2.0–200.0	81.8–107.5
	IL	Environmental water	Pesticides	HPLC-UV	0.5	1.5	2–250	58.9–85.8
	CNT	Environmental water	Pesticides	GC-MS	0.51–2.29	1.19–5.35	0.9040–0.9733	79.9–111.6
	Fe$_3$O$_4$@MAIDB	Environmental water	Triazole fungicides	HPLC-DAD	0.0050–0.0078	0.017–0.026	0.05–200.0	75.1–120
	Fe$_3$O$_4$-EDTA@Zr(IV)	Environmental water	OPPs	GC-MS	0.1–10.30	0.33–34.33	100–20000	86.95–112.60
	G@SiO$_2$@Fe$_3$O$_4$	Environmental water	OPPs	GC-FPD	16–33	50–100	50–5000	90.2–102.9
	Fe$_3$O$_4$@SiO$_2$-G	Environmental water	OCPs	GC-µECD	0.00012–0.00028	—	0.001–0.100	80.8–106.3
	MI-ILMM	Environmental water	OCPs	GC-µECD	0.12–0.26	0.40–0.87	1.0–100	86.2–100.4
	Co/HPC	Environment water and white gourd	Trizine herbicides	HPLC-DAD	0.02	—	0.2–100	80.3–120.6
	MOF	Drinking and environmental water	OCPs	GC-MS	0.00039–0.00070	0.00145–0.002	0.002–0.500	79.4–98.3

(Continued)

Table 5.7 (Continued)

Extraction technique	Adsorbent/ extraction solvent	Matrix	Analyte	Analytical technology	LOD (μg L^{-1})	LOQ (μg L^{-1})	LR (μg L^{-1})	R (%)
	Fe$_3$O$_4$@G-TEOS-MTMOS	Water	Polar and nonpolar OPPs	GC-μECD	0.0014–0.0237	0.0047–0.0789	0.1–1	83.09–110.53
	Fe$_3$O$_4$/SiO$_2$/CMCD	Water	Quaternary ammonium herbicides	CE-DAD	0.8–0.9	—	5–500	70.2–100.0
	Fe$_3$O$_4$–SiO$_2$–P4VP	Water	Phenoxy acid herbicide	CE-UV	3–10	10–30	10–500	80–85
	Magneto-liposome	Surface water of the Songhua River	OCPs	GC-MS/MS	0.00004–0.00035	0.00014–0.00059	0.0002–0.125	80–109
	Fe$_3$O$_4$@SiO$_2$@GO-PEA	Fruit, vegetable, and water	OPPs	GC-NPD	0.02–0.1	0.06–0.3	0.06–200	94.6–104.2
	IL	Water and fruit juices	OPPs	GC-FID	0.02–0.06	—	80–300000	70–89.2
	ZMNIC	Water and juice	OPPs	GC-MS	0.031–0.034	0.0105–0.112	0.1–500	86–106
	PPDA-co-Th@Fe$_3$O$_4$	Environmental water and fruits juice	OPPs	GC-FID	0.1–0.3	—	0.3–500	88.1–99.2
	KHA/Fe$_3$O$_4$	Environmental water and fruits juice	OPPs	GC-FID	0.03–0.22	—	0.1–200	89.0–99.7
	RGO/Fe$_3$O$_4$@Au	Seawater	OCPs	GC-ECD	0.0004–0.0041	0.0013–0.0136	0.05–500	69–114
	CoFe$_2$O$_4$-PGC Fe$_3$O$_4$–NH$_2$@MIL–101(Cr)	Environmental water	Pyrethroids	HPLC-DAD	—	—	—	80.2–110.9
		Environmental water	—	GC-ECD	0.005–0.009	0.015–0.025	0.020–2	72.1–106.8

M-MWCNTs	Environmental water	Triazole fungicides	HPLC-MS/MS	0.011–0.067	0.037 0.221	0.04–500.0	91.2–108.4
Carbon nanosphere @ Fe$_3$O$_4$	Environmental water	Triazole fungicides	HPLC-MS/MS	—	0.00056–0.00695	0.95	77.8–93.5
MMP/ZIF-8	Environmental water	Triazole fungicides	GC-MS/MS	0.08–0.27	—	1–400	83.4–96.9
MIL–101(Fe)@PDA@Fe$_3$O$_4$	Environmental water and vegetable Genuine water	Sulfonylurea herbicides	HPLC-DAD	0.12–0.34	—	1–150	87.1–108.9
Fe$_3$O$_4$@HPAMAM		Benzoylurea insecticides		0.39–0.72	2.5	2.5–500.0	75.1–111.4
MIL–101-NH$_2$@Fe$_3$O$_4$–COOH	Environmental water	Fungicides	HPLC-UV	0.04–0.4	—	0.2–100	71.1–99.1
MHMS-MCNP	Surface, ground environmental and wastewater	OPPs	HPLC-UV	0.07	0.23	0.2–1000	96–99
Fe$_3$O$_4$–NH$_2$–CAP	Water and soil	Phenylurea herbicides	HPLC-UV	0.01–0.03	0.03–0.09	0.10–40	6.0–103.0
3D Co/Ni@C	Water	Pyrethroids	HPLC-UV	0.0038–0.0067	0.013–0.022	0.1–100	85.6–106.9
Fe$_3$O$_4$@SiO$_2$@KIT–6	Environmental water	—	HPLC-UV	0.005–0.01	0.018–0.04	0.02–1200	86.58–98.80
TPN/Fe$_3$O$_4$ NPs/GO	Water and food	Acidic and basic pesticides	HPLC-UV	0.17–1.7	0.5–5.0	0.5–500	92.7–95.3
Fe$_3$O$_4$@SiO$_2$-GO-MOF	Environmental water	Benzenoid-containing insecticide	HPLC-UV	0.30–1.58	1.0–5.2	20.0–1000	81.2–113.1
Fe$_3$O$_4$–NH$_2$@MOF–235	Honey, fruit juice and tap water	Benzoylurea insecticides	HPLC-UV	0.25–0.5	—	1.0–300.0	84.02–99.62

(*Continued*)

Table 5.7 (Continued)

Extraction technique	Adsorbent/ extraction solvent	Matrix	Analyte	Analytical technology	LOD (µg L^{-1})	LOQ (µg L^{-1})	LR (µg L^{-1})	R (%)
	Fe$_3$O$_4$@PDA-DES	Environment and drinking water	Sulfonylurea herbicides	HPLC-PDA	0.0098–0.0110	—	5.0–200	72.5–108.6
	Fe$_3$O$_4$@CoFe$_2$O$_4$@PPy	Soil and water	Chlorophenoxy herbicides	HPLC-MS/MS	0.03–3.36	—	0.1–200	80–117
SPME	Carbonized PDA	Environmental water	OCPs	GC-MS	0.0014–0.015	—	0.05–50	72–109
	AuNPs-based SPME coating	Environmental water	OCPs	GC-ECD	0.13–0.24	0.44–0.81	—	85.0–97.1
	MIP	Environmental water	Triazines	HPLC-DAD	0.024–0.030 2.0–5.3	—	—	78–104
	Fe$_3$O$_4$@MIL–100 (Fe)	Environmental water	Triazine Herbicides	GC-ECD	—	6.1–15.7	0.0061–70	97.5–103
	Biosorbent-based fiber	Water	OCPs	GC-ECD	0.00019–0.00071	0.00065–0.00238	0.005–0.100	60–113
	Cork fiber	Water	OCPs	GC-ECD	0.0003–0.003	0.001–0.01	0.01–0.075	60.3–112.7
	C18	Water	OCPs	GC-MS	0.000059–0.000151	0.000197–0.000503	0.002–0.5	70.2–119
	Hierarchial G	Water	OCPs	GC-MS	0.00008–0.00080	0.00025–0.0027	0.01–30	82.8–113
	Mesoporous TiO$_2$ NP	Water	OCPs	GC-MS	0.00008–0.00060	0.00027–0.002	0.005–0.500	81.2–118.1
	Sol-gel/nanoclay	Water	OPPs	GC-MS	0.003–0.012	0.01–0.02	0.01–2.0	86–104
	Mesoporous silica	Water	Triazole pesticide	GC-MS	0.05–0.09	—	0.1–2000	—
	—	Water	Pesticide carbaryl	UV–VIS	0.033	0.11	100–1200	97–112
	PDMS/DVB	Surface and ground water	Pesticides	GC-MS	0.05–1	—	0.05–20	—
	CAR-PDMS	Waste water	Hydrazine	GC-MS/MS	0.002	0.007	0.9986	—
	MMF-MAED	Water and juice	Benzoylurea insecticides	HPLC-DAD	0.026–0.075	0.084–0.25	0.1–200	65.1–118

rGOQD@Fe	Fruit juice and real water	OPPs	GC-MS	0.04–0.07	0.11–0.21	0.14–540	82.9–113.2
CNT@SiO$_2$	Vegetables, fruits and water		GC-IMS	0.005–0.020	0.010–0.050	0.01–3	79–99
NH$_2$-MIL-53 (Al)	Tap and river water	OCPs	GC-MS	0.000051–0.00097	0.083–3.23	0.001–1	77.4–117
ZnO/g-C$_3$N$_4$	Cucumber, pear, Green tea and Minjiang water (local river)	Pesticides	GC-MS	0.001–0.0025	0.003–0.005	0.003–5.0	79.1–103.5
PA	River, fountain, reservoir and wellspring	Fungicides	GC-MS/MS	lower than 0.001	—	0.001–2	92–104
Co/Cr(NO$_3$)-LDH	Environmental water	Acidic pesticides	HPLC-UV	0.05–0.08		0.1–500	90–110
MMF	Environmental water and orange juice	Carbamate pesticide	HPLC-DAD	0.017–0.29	0.057–0.96	0.1–200	80.4–117
	Tap, river and waste waters	Sulfonylurea herbicides	HPLC-DAD	0.009–0.018	0.028–0.060	0.0500–200	70.1–108
	Water and soil	Sulfonylurea herbicides	HPLC-DAD	0.018–0.17	0.061–0.44	0.1–200.0	70.6–119
MOF	Environmental water	Fungicides	HPLC-MS	0.00134–0.0148	0.00445–0.0494	0.005–5	90.4–97.5
	Environmental water	Triazine herbicides	DART-MS	0.005–0.05	—	0.991	92.4–125.7
	Environmental water	OPPs	GC-NPD	0.005–0.008	0.015–0.025	0.015–50	88–108
PDMS	Surface water	Pesticides	GC-MS/MS	0.001–0.458	0.039–0.732	0.05–100	75.6–137.31
Si-GO	Ground and mineral water	Triazines	HPLC-MS/MS	0.0011–0.0029	0.0038–0.0097	0.2–4.0	—
MMC	Environmental water	Triazoles	HPLC-DAD	0.014–0.031	0.11–0.074	0.100–200	78.9–106
Magnetism-enhanced monolith	Environmental water	Triazines	HPLC-DAD	0.074–0.23	0.24–0.68	0.5–200	70.7–119
POM-IL	Environmental water	Triazole pesticides	GC–MS	0.005–0.03	—	0.02–100	—

(*Continued*)

Table 5.7 (Continued)

Extraction technique	Adsorbent/ extraction solvent	Matrix	Analyte	Analytical technology	LOD (µg L^{-1})	LOQ (µg L^{-1})	LR (µg L^{-1})	R (%)
SBSE	PDMS	Water	OCPs	GC–MS	0.00012–0.00207	—	0.01–0.5	64.7–111
	MWCNTs-PDMS	Water and soil	Triazine herbicide	IMS	0.006–0.015	0.02–0.05	0.05–10	85.9–104.3
	PDMS-MIL 100-Fe	Environmental water	Triazine	HPLC-UV	0.021–0.079	—	0.2–500	98–118
	ZnS-AC	Tap, river and mineral water	Carbamate insecticide		0.03–0.05	1–1.7	2–30000	91.7–98
	MOF	Environmental water	OPPs	GC-FPD	0.043–0.085	—	0.2–100	89.3–115
	CDD	Farm field water	Endocrine disruptor pesticides	GC-µECD	0.004–0.0045	0.001–0.015	0.005–5	83.2–98.7
	MOF	Soil and lake water	Sulfonylurea herbicides	HPLC-UV	0.04–0.84	—	10–800	68.8–98.1
	ZrO$_2$-rGO	River water and agricultural wastewater	Ethion	IMS	1.5	5.0	5.0–200	93–97
	MMT/PPy/N6	River	OPPs	GC-MS	0.05–0.3	—	0.3–1000	80.3–95.3
	ZrO$_2$/SiO$_2$ composite	Environmental and Food	OCPs	GC-MS	0.03–1.3	0.1–4	0.0001–5	69.0–114.1
	SMNM	Environmental water	OPPs insecticides	HPLC-UV	0.07–0.89	0.23–2.94	1–1000	80.60–104.52
LLE	Acetonitrile and DBE	Aqueous samples	Pesticides	GC-FID	0.34–5	1–16	1–10000	91–105
	DCM	Fresh water sediment	OTP	HPLC-MS	0.13–44	0.43–1.46	5–250	73.7–119.6
	DCM	Water	Benzulfuron-methyl herbicide	GC-MS	0.1	—	50–5000	74.1–94.1
	Salting-out solution (NaCl in	High salinity and biological samples	Pyrethroid insecticides	GC-MS	1.5–60.6	—	5–5000	74–110

	phosphate buffer) DCM and ethylacetate	Surface water, sediment, and fish	Pyrethroid insecticides	HPLC-PDA	0.01–0.03	0.05–0.07	0.07–0.49	97–99
	n-Hexane	Surface waters	Pyrethroid and OPPs insecticides	GC-MS	—	0.0125–0.125	0.98	67–114
	DCM and hexane	River, pond and tube well water and river sediment	Pesticides	GC-MS/MS	—	0.01–0.08	5–250	70.76–111.52
LPME	Nanofluids	Environmental water	Fungicides insecticides	HPLC-UV	0.13–0.19	0.44–0.64	0.996	74.94–96.11
	[C4MIM]PF6	Water	OPPs	HPLC-UV	0.01–0.1	0.05–0.4	0.09–200	96.9–103.2
	Decanoic acid and tetrabutyl-ammonium hydroxide	Water and fruit juice	OPPs	HPLC-UV	0.1–0.35	—	0.5–400	92.2–110.5
	1-Dodecanol/p-xylene	East lake, well and rainwater	OPPs	GC-FPD	0.12–0.56	—	0.6–100	83.7–112
	DCM and DCE	Dam lake, river and well water	Herbicides	GC-MS	0.3–2.0	6.5	5.0–100	83–121
DLLME	[HMIM]NTf2	Environmental water	Fungicides	HPLC-UV	0.02–0.10	—	0.5–500	70.7–105
	[P4448] [Br] and Na[N(CN)2]	Environmental water	Pyrethroid pesticides	HPLC-UV	0.16–0.21	—	1–100	80.20–117.31
	[BeMIM][Tf2N]	Environmental water	OPPs	HPLC-UV	0.01–1.0	—	0.9994–0.9998	82.7–118.3
	Acetic acid-containing CLF	Environmental water	OPPs	GC-FPD	0.004–0.03	—	0.03–300	75.8–104.2
	1-Octanol	Environmental water	Multiresidue pesticides	HPLC-DAD	0.8–3.3	2.5–11.0	8.6–800	81–121

(*Continued*)

Table 5.7 (Continued)

Extraction technique	Adsorbent/ extraction solvent	Matrix	Analyte	Analytical technology	LOD (μg L^{-1})	LOQ (μg L^{-1})	LR (μg L^{-1})	R (%)
	TCE	Water	Pesticide	GC-MS	0.0032–0.0174	0.0096–0.052	0.0096–100	84–108
	([C$_4$MIm] Cl) and (Li [NTf$_2$])	Water	OPPs	GC-MS	0.005–0.16	0.017–0.054	0.005–0.016	97–113
	M-β-CD/ATP	Water	Fungicides	HPLC-DAD	0.02–0.04	—	1.00–500	98–115
	[P$_{44412}$][PF$_6$]	Water	Pyrethroid insecticides	HPLC-UV	0.71–1.54	—	1–500	87.1–101.7
	Decanoic acid	Water	Pesticide	HPLC-UV	0.24–0.68	—	1–500	84.7–95.3
	1-Octanol	Water		HPLC-MS/MS	—	0.0125–1.25	0.9849–0.9962	60–120
	LiNTf$_2$	Water	Triazole fungicide	HPLC-UV	0.74–1.44	—	5–250	70.1–115
	1-Undecanol	Water and grape juice	Herbicides and fungicides	HPLC-MS	0.0027–0.0097	0.009–0.0323	0.05–10	72.4–101.5
	[N8881] PF$_6$	Water and tea beverage	Benzoylurea insecticides	HPLC-UV	0.29–0.59	0.97–1.97	2–500	85.93–90.52
	DBE	River water and fruit juice	OPPs	GC-FID	0.82–2.72	2.60–7.36	3.1–1000	64–83
	1-Dodecanol	Mineral water	Triazine and triazoles	HPLC-MS	0.00375–0.0375	0.0125–0.125	0.125–3.75	70–118
	CLF and MCB	Mineral water	Triazine, neonicotinoid triazole and imidazolinone pesticide	HPLC-MS/MS	—	0.005–0.5	0.5–15	102–120
	CTC	Water and soil	Fungicide carbendazim	UV–Vis	2.1	4	5–600	95.2–105.2
	CLF and MeOH	Tap and waste water	Pesticide	GC-MS	0.09–3.36	0.31–11.19	0.50–500	84–109
	Aliquat–336 in 1-octanol	Water and soil	Sulfonyl urea herbicide	HPLC-UV	0.5	—	1–100	89.8–110.1
	DBE	Aqueous sample and Fruit juice	OPPs	GC-FID	0.65–1.3	2.2–4.5	3–40000	67–95

Tetrachloro-ethene	Sea and surface water	OPPs	GC-MS/MS	0.003	0.007–0.012	—	30–130
Tetrachloro-ethene	Aqueous samples, fruit juices, and vegetables	OPPs	GC-NPD	0.012–0.056	0.044–0.162	0.05–60	84–92
CLF and undecanol	Water and fruit samples	Fungicides	HPLC-MS/MS	—	0.001	—	86–116
DCE	Tap water and wastewater	Pesticides	GC-MS	0.30–2.0	0.53–5.4	0.50–1000	86–107
Decane	Honey, tomato and environ-mental water	Triazine herbicide	S-MEKC	0.07–0.69	—	0.3–100	86.4–114
CLF and DCM	Natural water	Pesticides	HPLC-DAD	0.015–0.36	0.049–1.2	2.9–2500	84.8–106.1
[C$_6$MIM][PF$_6$]	Surface sample		HPLC-MS/MS	0.1–0.8	0.5–2.5	0.5–50	70–120
Trichloro-ethylene	Environmental water	Herbicides	GC-MS	0.045–0.03	0.10–0.15	0.10–200	87.2–111.2
CLF and DCE	Wastewater and sludge	OCPs		0.16–1.5	0.52–4.9	0.50–500	82–108
1-Dodecanol	Environmental water	Triazole fungicides	HPLC-DAD	5.1–9	16.9–29.9	50–5000	77.6–104.4
ChCl(ASA)$_2$ MeOH	Surface water	Fungicides, insecticides and acaricides	HPLC-MS/MS	0.002–2.3	0.006–7.7	0.9737–0.9975	18–96
CLF	Waters, soil, food and beverage	Carbamate insecticides	IMS	1.04–1.31	—	2–100	96–101
DES	Environmental water	Pyrethroid insecticides	HPLC-UV	0.30–0.60	1.00–2.00	0.9992–0.9995	80.93–109.88
LiNTf$_2$	Environmental water	Fungicides	HPLC-DAD	0.032–0.89	—	1–100	89.4–96.4
[MimCH$_2$COOCH$_3$][NTf$_2$]	Environmental water	OPPs	HPLC-UV	0.7–2.7	—	1–100	96.3–114.2

(Continued)

Table 5.7 (Continued)

Extraction technique	Adsorbent/extraction solvent	Matrix	Analyte	Analytical technology	LOD (μg L^{-1})	LOQ (μg L^{-1})	LR (μg L^{-1})	R (%)
	Toluene	Environmental water	Triazole fungicides	GC-MS	0.14–0.27	0.47–0.90	1–100	89.3–108.7
	1-Dodecanol	Environmental water	Pyrethroid	HPLC-VWD	0.37–0.75	—	1–500	91.3–98.1
SPE-DLLME	MWCNT, silica-based C18, and polymeric solid sorbent	Environmental water	OCP	GC-MS	0.005–0.22	—	500	—
	MIP	Tap water, ground water, grape juice	Triazine herbicides	GC-FID	0.2–7	0.5–20	0.5–150	92–98
	C18	Water, milk, honey and fruit juice	Pesticide	GC-MS	0.0005–0.001	—	0.001–10	78.1–105
MSPE-DLLME	MOF	Environmental water and tea	Pyrethroids	GC-ECD	0.008–0.015	0.028–0.050	0.05–10.0	78.3–103.6
	MGO	Environmental water	Atrazine herbicide	GC-MS	0.0006	0.002	0.005–5	96–102
QuEChERS		Water	Pesticide	GC-MS	0.3–4	0.95–13.69	0.995–0.999	85.3–107
		Soil and water	OCPs	GC-MS	0.26–1.89	0.86–9.99	0.9931–0.9999	70.0–118.0
		Food and water	Pesticides	HPLC-MS/MS	—	0.0001–0.0478	0.994–0.999	93–96

Main abbreviations (for techniques and pesticide classes): *DLLME*, dispersive liquid-liquid microextraction; *LPME*, liquid-phase microextraction; *LLE*, liquid-liquid extraction; *OCP*, organochlorine pesticide; *OPP*, organophosphorus pesticide; *OTP*, organotin pesticide; *QuEChERS*, quick, easy, cheap, effective, rugged, and safe; *SBSE*, stir bar sorptive extraction; *SPE*, solid-phase extraction; *SPME*, solid-phase microextraction.

Source: Adapted from Nasiri, M., Ahmadzadeh, H., & Amiri, A. (2020). Sample preparation and extraction methods for pesticides in aquatic environments: A review. *TrAC, Trends in Analytical Chemistry*, *123*, 115772. Reprinted with permission from Elsevier.

Main analytical techniques   167

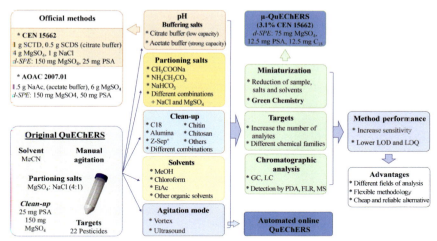

Figure 5.23 Relevant improvements in QuEChERS protocol. *d-SPE*, dispersive solid-phase extraction; *EtAc*, ethyl acetate; *FLR*, fluorescence detector; *GC*, gas chromatography; *LC*, liquid chromatography; *LOD*, limit of detection; *LOQ*, limit of quantification; *MeCN*, acetonitrile; *MeOH*, methanol; *MS*, mass spectrometry; *PDA*, photodiode array detector; *PSA*, primary secondary amine; *SCDS*, Sodium citrate dibasic sesquihydrate; *SCTD*, sodium citrate tribasic dihydrate; *Z-Sep*, Zirconium dioxide based sorbents. *Reproduced with permission from Elsevier, Perestrelo, R., Silva, P., Porto-Figueira, P., Pereira, J. A. M., Silva, C., Medina, S., & Câmara, J. S. (2019). QuEChERS—fundamentals, relevant improvements, applications and future trends.* Analytica Chimica Acta, 1070, 1–28.

(SPME), liquid-phase microextraction (LPME) and dispersive liquid–liquid microextraction (DLLME), and variations of each ones (Nasiri, Ahmadzadeh, & Amiri, 2020).

The QuEChERS extraction technology should be highlighted for residue analysis in food due the fact that it contributes to achieve lower LOD and LOQ from the analyte recovery. Perestrelo et al. (2019) describe the QuEChERS approach (Fig. 5.23). As cited before, Shabeer et al. (2018) developed and reported for the first time a comprehensive multiresidue GC-MS/MS method for 243 pesticides in cardamom associated to QuEChERS extraction.

5.9 Conclusion

As seen in this chapter, analytical techniques play an important role in the analysis of agrochemical residues in food and in the environment, giving the knowledge of chemical composition and presence or absence of contaminants and pollutants.

Nowadays, there are a large number of analytical techniques available to the laboratories according their necessities based on samples (matrices and analytes). For agricultural purposes we can highlight spectroscopic and spectrometric techniques (e.g., UV–vis, FTIR, NIR, AAS and OES); chromatographic techniques (liquid and gaseous phases) hyphenated to a large variety of detectors; thermal analysis and microscopy. Moreover, sensor and probes has been gaining more and more space due to its ease of handling and the speed in generating results.

However, extraction step deserve a special attention in order to guarantee the reliability of the analytical results.

References

Banica, F.-G. (2012). *Chemical sensors and biosensors: Fundamentals and applications*. Chichester: Wiley.

Chawla, P., Kawshik, R., Swaraj, V. J. S., & Kumar, N. (2018). Organophosphorus pesticides residues in food and their colorimetric detection. *Environmental Nanotechnology, Monitoring & Management*, *10*, 292–307.

Dehotay, S. J., & Cook, J. M. (2015). Sampling and sample processing in pesticide residue analysis. *Journal of Agricultural and Food Chemistry*, *63*, 4395–4404.

EAG Laboratories. (2020). Techniques. <https://www.eag.com/> accessed 10.20.

Harris, D. C. (2010). *Quantitative chemical analysis* (p. 719) W. H. Freeman, ISBN 9781429218153.

Hernández-Mesa, M., Moreno-González, D., Lara, F. J., & García-Campaña, A. M. (2019). Electrophoresis | capillary electrophoresis: Food chemistry applications. *Encyclopedia of analytical science* (3rd ed., pp. 358–366). Amsterdam: Elsevier.

Hou, X., Xu, X., Xu, X., Han, M., & Qiu, S. (2020). Application of a multiclass screening method for veterinary drugs and pesticides using HPLC-QTOF-MS in egg samples. *Food Chemistry*, *309*, 125746.

Hurtado-Sánchez, M., del, C., Lozano, V. A., Rodríguez-Cáceres, M. I., Durán-Merás, I., & Escandar, G. M. (2015). Green analytical determination of emerging pollutants in environmental waters using excitation–emission photoinduced fluorescence data and multivariate calibration. *Talanta*, *134*, 215–223.

International Union of Pure and Applied Chemistry. IUPAC. (2020). IUPAC compendium of chemical terminology—the gold book. <http://goldbook.iupac.org/index.html> Accessed 10.20.

Jin, W., & Maduraiveeran, G. (2017). Electrochemical detection of chemical pollutants based on gold nanomaterials. *TrEAC, Trends in Environmental Analytical Chemistry*, *14*, 28–36.

Jørgensen, N., Laursen, J., Viksna, A., Pind, N., & Holm, P. E. (2005). Multi-elemental EDXRF mapping of polluted soil from former horticultural land. *Environment International*, *31*, 43–52.

Li, Y., Luo, Q., Hu, R., & Chen, Z. (2018). A sensitive and rapid UV–vis spectrophotometry for organophosphorus pesticides detection based on ytterbium (Yb^{3+}) functionalized gold nanoparticle. *Chinese Chemical Letters*, *29*, 1845–1848.

Liana, D. D., Raguse, B., Gooding, J. J., & Chow, E. (2012). Recent advances in paper-based sensors. *Sensors, 12*, 11505−11526.

Meredith, N. A., Quinn, C., Cate, D. M., Ill, T. H. R., Volckens, J., & Henry, C. H. (2016). Paper-based analytical devices for environmental analysis. *Analyst, 141*, 1874−1887.

Miner, G. L., Delgado, J. A., Ippolito, J. A., Barbarick, K. A., Stewart, C. E., Manter, D. A., ... D'Adamo, R. E. (2018). Influence of long-term nitrogen fertilization on crop and soil micronutrients in a no-till maize cropping system. *Field Crops Research, 228*, 170−182.

Mitra, S. (Ed.), (2003). *Sample preparation techniques in analytical chemistry*. Hoboken: Wiley.

Nasiri, M., Ahmadzadeh, H., & Amiri. (2020). Sample preparation and extraction methods for pesticides in aquatic environments: A review. *TrAC, Trends in Analytical Chemistry, 123*, 115772.

Pagani, A. P., & Ibáñez, G. A. (2019). Pesticide residues in fruits and vegetables: High-order calibration based on spectrofluorimetric/pH data. *Microchemical Journal, 149*, 104042.

Perestrelo, R., Silva, P., Porto-Figueira, P., Pereira, J. A. M., Silva, C., Medina, S., & Câmara, J. S. (2019). QuEChERS—fundamentals, relevant improvements, applications and future trends, . *Analytica Chimica Acta* (1070, pp. 1−28). .

Shabeer, T. P. A., Girame, R., Utture, S., Oulkar, D., Banerjee, K., Ajay, D., ... Menon, K. R. K. (2018). *Optimization of multi-residue method for targeted screening and quantitation of 243 pesticide residues in cardamom (*Elettaria cardamomum*) by gas chromatography tandem mass spectrometry (GC-MS/MS) analysis*, . *Chemosphere* (193, pp. 447−453). .

Skoog, D. A., West, D. M., Crouch, S. R., & Holler, F. J. (2014). *Fundamentals of analytical chemistry* (p. 958) Brooks/Cole, Cengage Learning, ISBN 9781285056241.

U.S. Environmental Protection Agency. (1997). Standard operation procedure—soil sampling. New York: EPA.

Vaz, S., Jr (2018). *Analytical techniques. Analytical chemistry applied to emerging pollutants*. Cham: Springer Nature.

Wieczerzak, M., Namiésnik, J., & Budlak, B. (2016). Bioassays as one of the green chemistry tools for assessing environmental quality: A review. *Environmental International, 94*, 341−361.

Zeng, X., Zhang, Y., Zhang, J., Hu, H., Wu, X., Long, Z., & Hou, X. (2017). Facile colorimetric sensing of Pb^{2+} using bimetallic lanthanide metal-organic frameworks as luminescent probe for field screen analysis of lead-polluted environmental water. *Microchemical Journal, 134*, 140−145.

Zhang, C., Khang, K., Zhao, T., Liu, B., Wang, Z., & Zhang, Z. (2017). Selective phosphorescence sensing of pesticide based on the inhibition of silver(I) quenched ZnS:Mn^{2+} quantum dots. *Sensors and Actuators B: Chemical, 252*, 1083−1088.

CHAPTER 6

Analytical methods to selected matrices

Analytical methods represent the functionality of the analytical techniques and the analytical chemistry fundamentals. They will use theoretical concepts (e.g., the absorption and emission of radiation) behind techniques considering figures of merit in order to construct the better approach to solve a certain problem, that is, the determination (quantitative or qualitative) of an analyte in a matrix.

When we consider food and environmental-related matrices it is paramount that methods be robust[1] because the first are high-heterogeneous chemical mediums of analyzes. Moreover, multiresidue methods are preferable instead of one-single method. Additionally, special attention should be given to the use of certified reference materials (CRMs) to validate the method in order to guarantee its quality control (Chapter 4: Fundamentals of analytical chemistry).

A reduction in the analysis steps, in energy and reactants consumption, and in the residue generation will provide a desirable green method.

6.1 Analytical method for organic residues (pesticides) in the environment

Organochroline pesticides are one of the most representative agrochemical classes used in the modern agriculture—for instance, as insecticides. Its use is intended to make agricultural production viable through the extermination of pests, which often end up with food production. However, after their use, these compounds remain active in the environment for long periods, damaging ecosystems—that means, they are persistent organic pollutants (POPs).

---

[1] Robustness has been defined as being the capacity of an analytical procedure to produce unbiased results when small changes in the experimental conditions are made voluntarily. For more information: https://doi.org/10.1016/j.microc.2016.12.004

6.1.1 Method
Organochlorine pesticides by gas chromatography.

6.1.2 Source
SW-846[2] test method 8081B (U.S. Environmental Protection Agency, 2007)—free of charge.

6.1.3 Scope and application
This method may be used to determine the concentrations of various organochlorine pesticides in extracts from solid and liquid matrices, using fused-silica, open-tubular, capillary columns with electron capture detectors (ECD) or electrolytic conductivity detectors (ELCD). The following compounds—and their CAS number—may also be determined using this method: Alachlor (15972-60-8), Captafol (2425-06-1), Carbophenothion (786-19-6), Chloroneb (2675-77-6), Chloropropylate (5836-10-2), Chlorothalonil (1897-45-6), Dacthal (DCPA) (1861-32-1), Dichlone (117-80-6), Dichloran (99-30-9), Dicofol (115-32-2), Etridiazole (2593-15-9), Halowax-1000 (58718-66-4), Halowax-1001 (58718-67-5), Halowax-1013 (12616-35-2), Halowax-1014 (12616-36-3), Halowax-1051 (2234-13-1), Halowax-1099 (39450-05-0), Mirex (2385-85-5), Nitrofen (1836-75-5), *trans*-Nonachlor (39765-80-5), Pentachloronitrobenzene (PCNB) (82-68-8), Permethrin (*cis* + *trans*) (52645-53-1), Perthane (72-56-0), Propachlor (1918-16-7), Strobane (8001-50-1), Trifluralin (1582-09-8).

6.1.4 Description
A measured volume or weight of liquid or solid sample is extracted using the appropriate matrix-specific sample extraction technique. After that:
1. Aqueous samples may be extracted at neutral pH value with methylene chloride using either Method 3510 (separatory funnel), Method 3520 (continuous liquid-liquid extractor), Method 3535 (solid-phase extraction), or other appropriate technique—additional methods available on SW-846 Compendium.
2. Solid samples may be extracted with hexane-acetone (1:1) or methylene chloride-acetone (1:1) using Method 3540 (Soxhlet), Method 3541 (automated Soxhlet), Method 3545 (pressurized fluid extraction), Method 3546 (microwave extraction), Method 3550 (ultrasonic

---
[2] The SW-846 Compendium: https://www.epa.gov/hw-sw846/sw-846-compendium

extraction), Method 3562 (supercritical fluid extraction), or other appropriate technique or solvents—again, additional methods available on SW-846 Compendium.
3. A variety of cleanup steps may be applied to the extract, depending on the nature of the matrix interferences and the target analytes. Suggested cleanups include alumina (Method 3610), florisil (Method 3620), silica gel (Method 3630), gel permeation chromatography (Method 3640), and sulfur (Method 3660)—again, additional methods available on SW-846 Compendium.
4. After cleanup, the extract is analyzed by injecting a measured aliquot into a gas chromatograph equipped with either a narrow-bore or wide-bore fused-silica capillary column, or either an ECD (GC-ECD) (Fig. 6.1) or an ELCD (GC-ELCD).

Critical evaluation: *This is a very useful U.S.-EPA method for the determination of organochlorine residues in environmental matrices (i.e., water and soil). However, its application will need a solid understanding of requisites of quality assurance & quality control (QA/QC). A good knowledge about the equipment operation is desirable.*

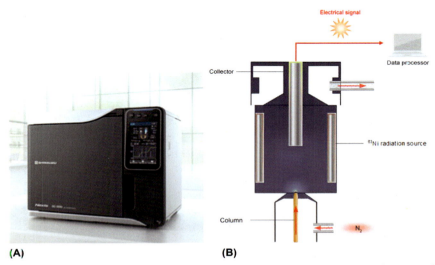

Figure 6.1 A gas chromatograph equipment (A, *left*), and the schematic diagram of the ECD (B, *right*). *Courtesy Shimadzu.*

174　Analysis of Chemical Residues in Agriculture

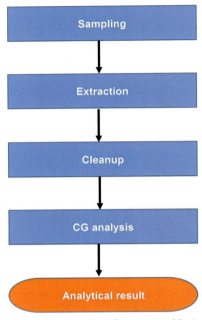

Figure 6.2 *Modus operandi* of the method for organochlorine residues in the environment. *Credit: author.*

Fig. 6.2 describes a simplified flowchart of the method.

6.2 Analytical method for organic residues (pesticides) in vegetables and fruits, and their wastes

Organophosphorous, carbamates, phenylureas, neonicotinoids, and triazines, are commonly used in fruit and vegetable crops in order to control insects and fungi.

6.2.1 Method

Multiresidue method for the analysis of 101 pesticides and their degradates by liquid chromatography-time-of-flight mass spectrometry.

6.2.2 Source

Ferrer and Thurman (2007).

6.2.3 Scope and application

A LC—time of flight mass spectrometry (LC-TOF-MS) method for multiresidue analysis of 101 pesticides (organophosphorous, carbamates, phenylureas, neonicotinoids and triazines, etc.) in vegetables and fruit samples is described. Different fruit and vegetable samples like green pepper, tomato, cucumber, and orange are subject of this analysis by using QuEchERS method of extraction (Fig. 6.3).

6.2.4 Description

15 g of homogenized sample is taken in a centrifuged tube, mixed with 15 mL of acetonitrile and vigorously shaken for 1 minute. 1.5 g of NaCl and 4 g of $MgSO_4$ are added to the mixture and shaken again for 1 minute. The solution is centrifuged at 3700 rpm for 1 minute. 5 mL aliquot supernatant is taken in another centrifuged tube containing 250 mg of PSA and 750 mg of $MgSO_4$ and was again shaken for 20 seconds. The extract was centrifuged for 1 minute at 3700 rpm. The extract is evaporated and reconstituted up to 1 mL and injected on LC—TOF-MS and

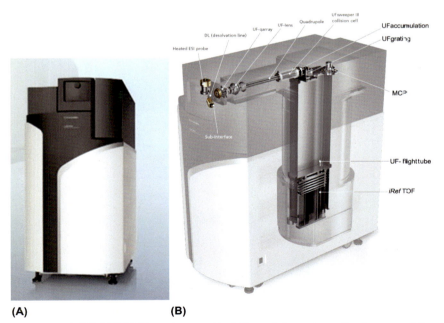

Figure 6.3 A HPLC-TOF-MS equipment (A, *left*), and the schematic diagram of the TOF-MS detector (B, *right*). *Courtesy Shimadzu.*

HPLC for the analysis of pesticides like dimethoate, malathion, chlorpyrifos methyl, captan, carbaryl, alachlor, aldicarb, acetamiprid, imazali, metalaxyl, nicosulfuron, trifluralin, propanil, thiosultap and simazine, etc.

The separation of the selected herbicides is carried out using an HPLC system (consisting of vacuum degasser, autosampler and a binary pump) equipped with a reversed phase C_8 analytical column of 150 mm × 4.6 mm and 5 μm particle size. Column temperature is maintained at 25°C. The injected sample volume is 50 μL. Mobile phases A and B are acetonitrile and water with 0.1% v/v formic acid, respectively. The optimized chromatographic method keeps the initial mobile phase composition (10% A) constant for 5 minutes, followed by a linear gradient to 100% A after 30 minutes. The flow-rate used is 0.6 mL min^{-1}. A 10 minutes post-run time was used after each analysis. This HPLC system is connected to a time-of-flight mass spectrometer (Fig. 6.4) equipped with an electrospray interface operating in positive ion mode, using the following operation parameters: capillary voltage, 4000 V; nebulizer pressure, 40 psig; drying gas, 9 L min^{-1}; gas temperature, 300°C; fragmentor voltage, 190 V; skimmer voltage, 60 V; octopole d.c. 1, 37.5 V; octopole RF, 250 V. LC-MS accurate mass spectra were recorded across the range 50–1000 m/z. The data recorded is processed with a software with accurate mass application-specific additions. Accurate mass measurements of each peak from the total

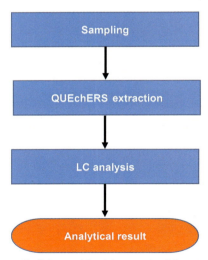

Figure 6.4 *Modus operandi* of the multiresidue method for pesticide residues in fruits and vegetables, and their wastes. *Credit: author.*

ion chromatograms is obtained by means of an automated calibrant delivery system using a dual-nebulizer ESI source that introduces the flow from the outlet of the chromatograph together with a low flow of a calibrating solution, which contains the internal reference masses—purine ($C_5H_4N_4$) at m/z 121.0509 and HP-921 [hexakis-(1H,1H,3H-tetrafluoro-pentoxy)phosphazene] ($C_{18}H_{18}O_6N_3P_3F_{24}$) at m/z 922.0098. The instrument should work providing a typical resolution of 9700 ± 500 (m/z 922).

Critical evaluation: *it is a high-sensitive method to determine a large number of pesticide residues in food matrices. However, its main inconvenience is the necessity of high investment to purchase the equipment what can be overcome if the laboratory has a high demand for this type of analysis.*

Fig. 6.4 illustrates a simplified flowchart for the method.

6.3 Analytical method for organic residue (veterinary drug) in meat

Benzimidazole is a veterinary drug that kills parasitic worms and it is used therapeutically in the treatment of helminthiasis in livestock. However, its residue could reach the muscle (meat).

6.3.1 Method
LC—MS/MS method for determination of benzimidazole residue in animal products.

6.3.2 Source
International Atomic Energy Agency (2016), Food and Environmental Protection Section.

6.3.3 Scope and application
This LC—ESI—MS/MS method is suitable for determination of benzimidazole (BZs), pro—benzimidazoles and their metabolites in animal products including pork, mutton, liver, milk, and fish. The target BZs, pro—benzimidazole, and their metabolites include 5—hydroxy—thiabendazole (TBZ—5—OH), thiabendazole (TBZ), albendazole—2—aminosulfone (ABZ—NH_2—SO_2), albendazole sulfoxide (ABZ—SO),

oxibendazole (OXI), oxfendazole (OXF), albendazole sulfone (ABZ—SO$_2$), albendazole (ABZ), febantel, thiophenate—ethyl, fenbendazole sulfone (FBZ—SO$_2$) and fenbendazole (FBZ). The LOD and LOQ are 0.75 and 2.5 µg/kg, respectively.

6.3.4 Description

Samples are extracted with potassium carbonate and ethyl acetate and defatted using hexane. Qualitative and quantitative measurement of the residues is done by LC—ESI—MS/MS with or without internal standard ($^{13}C_6$—thiabendazole).

The method comprises the follow steps:

1. Add 50 µg L^{-1} of the internal standard, 3 g of sodium sulfate, 3 mL 2 mol L^{-1} potassium carbonate and 15 mL of ethyl acetate to 5 g of homogenized muscle or liver tissue, in a 50 mL centrifuge tube;
2. Shake on a vortex mixer for 2 minutes and centrifuge for 5 minutes at 5000 rpm;
3. Decant the supernatant into 100 mL a distillation flask;
4. Repeat the extraction procedure once using 15 mL of ethyl acetate;
5. Evaporate the collected organic phases to dryness under nitrogen (at 45°C);
6. Add 3 mL MeCN and 5 mL *n*—hexane to the dried residue and shake for 2 minutes with an ultra—sonicator before transferring content into a 10 mL centrifuge tube;
7. Centrifuge for 5 minutes at 5000 rpm and, discard the upper layer;
8. Add 5 mL of *n*—hexane to the remaining layer to defeat the extract;
9. Evaporate the MeCN layer to dryness under nitrogen (at 45°C) and redissolve the residue in 1 mL of MeCN; pipet 100 µL (step i) solution into 900 µL MeCN:H$_2$O (30:70, v/v) and press the material through a 0.22 µm filter material;
10. Inject into an LC—MS/MS for analysis (Fig. 6.5).

Critical evaluation: *It is a method that provides a high-speed analysis of multiple residues with adequate robustness and ease of use. As for the item 6.2, it needs a high investment to purchase the equipment what can be overcome if the laboratory has a high demand for this type of analysis. Acetonitrile (MeCN)—a component of the mobile phase—is toxic by skin absorption and requires special care in its handling.*

Fig. 6.6 describes the simplified flowchart of the method.

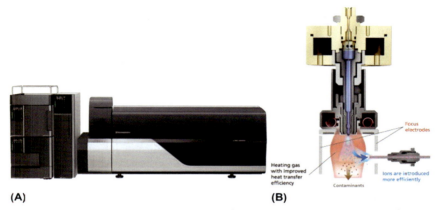

Figure 6.5 A HPLC-MS/MS equipment (A, *left*), and the schematic diagram of an ESI device (B, *right*). *Courtesy Shimadzu.*

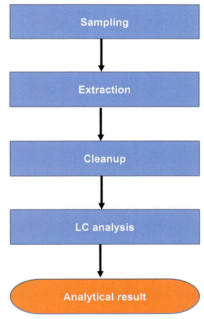

Figure 6.6 *Modus operandi* of the method for veterinary drug residue in muscle (meat). *Credit: author.*

6.4 Analytical method for organic residues (additives) in processed product (food, beverage, and feed)

According to the European Union (2020), feed additives can't be put on the market unless authorization has been given following a scientific evaluation demonstrating that the additive has no harmful effects, on human and animal health and on the environment. Then these chemicals should be carefully monitored and controlled.

For instance, dyes tartrazine (E102), sunset yellow FCF (E110), ponceau 4R (E124) are sulfonated azo dyes commonly added to food and beverages in order to improve the appearance and make them more attractive—they can be used for animal feed also. However, we can observe several indicatives of health problems from both uses, like frequent headaches in adults, neurotoxicity, genotoxicity, and carcinogenicity.

6.4.1 Method
A hybrid sorption–spectrometric method for determination of synthetic anionic dyes in food stuffs.

6.4.2 Source
Tikhomirova, Ramazanova, and Apyari (2017).

6.4.3 Scope and application
A sorption-spectrometric method for determination of the anionic synthetic dyes based on their sorption on silica sorbent modified with hexadecyl groups (C16) followed by measuring the diffuse reflectance spectra on the surface. The task of isolation of the dyes from complex matrices is actual and sorption approach can be the most promising way to solve this challenge.

6.4.4 Description
It is of interest to carry out detection of dyes directly on the adsorbent surface, for example, using diffuse reflectance spectroscopy. The apparent advantage of this method is reduction of the number of operations, which is associated with combining preconcentration of an analyte onto a solid form suitable for measurement and the measurement of the analytical response. The methodology is composed by the following steps:

1. Stock solutions of the dyes (concentration = 1 g L^{-1}) are prepared by dissolving a weighed portion of the reagent in distilled water. Working solutions (concentration = 0.1 g L^{-1}) is prepared by diluting the stock solutions in water. To adjust the desired pH values, 1 mol L^{-1} and 4 mol L^{-1} HCl and 0.01 mol L^{-1} NaOH is used. The ionic strength is adjusted by 2 mol L^{-1} and 4 mol L^{-1} NaCl. All reactants are at least of analytical grade.

2. Absorbances are measured by a spectrophotometer (UV-VIS region) (Fig. 6.7). Test-tubes with adsorbents are shaken on mechanical shaker. The pH values are measured using a pH meter, diffuse reflectance spectra are recorded by a mini-spectrophotometer—a monitor calibrator, as an effective alternative to a diffuse reflectance spectrometer. Manifold N 017 water jet pump is used during sorption of dyes in a dynamic mode.

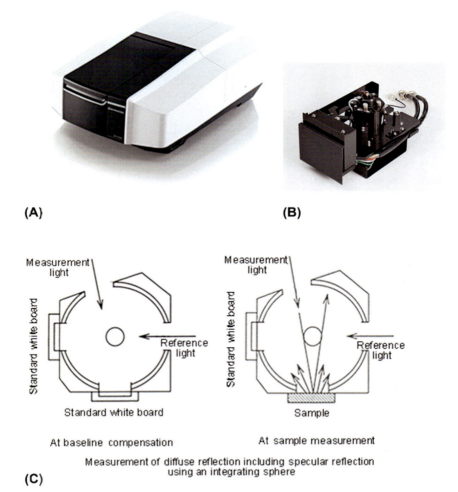

Figure 6.7 An UV-VIS spectrophotometer equipment (A, *left*) and an integrating sphere attachment device for the first (B, *right*); and the schematic diagram of diffuse reflectance (C, *left*). *Courtesy Shimadzu.*

3. Diasorb-130-C16 (C16) (specific surface area of 250 m² g⁻¹, particle size of 40–160 μm, pore diameter of 13 nm, total specific pore volume 1.00–1.125 cm³ g⁻¹) and KSK-G silica (specific surface of 250 m² g⁻¹, particle size of 40–160 μm, pore diameter of 10.5 nm) are used as the adsorbents.
4. To study the sorption in a static mode, 1 mL of a dye working solution (0.1 g L⁻¹) is introduced into a 15 mL graduated test-tube, HCl is added to adjust pH in the range of 1 mol L⁻¹—pH value of 7, then a portion of the C16 adsorbent of 0.10 g is introduced, and the mixture is diluted up to 10 mL with water. The modified silica sample is conditioned with a small amount of ethanol before addition into a solution. The test-tubes are shaken during 15 minutes for equilibration, then the adsorbent is separated from the solution by filtration on filter paper, and air dried. Then diffuse reflectance spectra is recorded. The content of the dye in a solution is determined spectrophotometrically by its own extinction at a wavelength (λ) of maximum absorbance ($\lambda_{Tartrazine}$ = 427 nm, $\lambda_{Sunset\ yellow}$ = 484 nm, $\lambda_{Ponceau\ 4R}$ = 506 nm).
5. When performing sorption in a dynamic mode, a column (diameter of 0.5 cm) packed with 0.2 g of the adsorbent is used. A sample solution is passed through the column at the rate of 0.5–1.0 mL min⁻¹ using a water jet pump.

Critical evaluation: As an alternative to HPLC analysis, this spectrophotometric-based method is cost-effective with an adequate laboratory performance. Moreover, it reduces steps and the quantity of solvents and reactants. However, matrix effects should be monitored and taken into account. Additionally, adsorption degree and selectivity of adsorption are pH- and ionic strength-dependent.

Fig. 6.8 describes the simplified flowchart of the method.

6.5 Analytical method for inorganic residues in the environment

The use of inorganic fertilizers can be a source of inorganic pollutants which can reach the groundwater.

6.5.1 Method

Multielemental determination by inductively coupled plasma-optical emission spectrometry (ICP-OES).

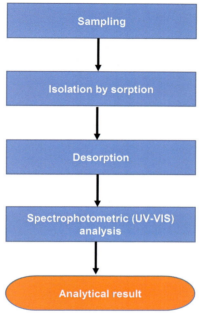

Figure 6.8 *Modus operandi* of the method for azo dyes in food and feed. *Credit: author.*

6.5.2 Source

SW-846 test method 6010D (U.S. Environmental Protection Agency, 2007).

6.5.3 Scope and application

ICP-OES is a spectrometric technique used to determine trace elements in aqueous solutions. In ICP-OES (Fig. 6.9), a sample solution is aspirated (i.e., nebulized) continuously into an inductively coupled, argon-plasma discharge, where analytes of interest are converted to excited-state, gas-phase atoms or ions. As the excited-state atoms or ions return to their ground state, they emit energy in the form of light at wavelengths that are characteristic of each specific element. The intensity of the energy emitted at the chosen wavelength is proportional to the amount (concentration) of that element in the analyzed sample. Thus, by determining which wavelengths are emitted by a sample and their respective intensities, the elemental composition of the given sample relative to a reference standard may be quantified. For accurate results, direct ICP-OES analysis

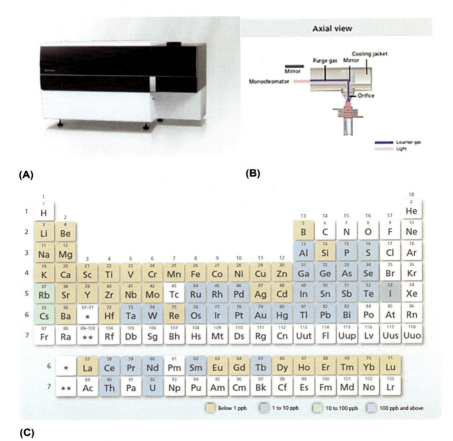

Figure 6.9 An ICP-OES equipment (A, left) and its internal torch system (B, right); for the analyzes of several elements (C, left). *Courtesy Shimadzu.*

should be conducted on only relatively clean, aqueous matrices (e.g., prefiltered groundwater samples). Other, more complex aqueous and/or solid samples need acid digestion prior to analysis; the analyst should ensure that a sample digestion method is chosen that is appropriate for each analyte and the intended use of the data. Refer to SW-846 for the appropriate digestion method.

The following analytes—and its CAS number—have been determined by this method: Ag (7440-22-4), Al (7429-90-5), As (7440-38-2), B (7440-42-8), Ba (7440-39-3), Be (7440-41-7), Ca (7440-70-2), Cd (7440-43-9), Co (7440-48-4), Cr (7440-47-3), Cu (7440-50-8), Fe (7439-89-6), Hg (7439-97-6), Li (7439-93-2), Mg (7439-95-4),

Mn (7439-96-5), Mo (7439-98-7), Na (7440-23-5), Ni (7440-02-0), P (7723-14-0), Pb (7439-92-1), K (7440-09-7), Sb (7440-36-0), Se (7782-49-2), Si (7631-86-9), Sn (7440-31-5), Sr (7440-24-6), Ti (7440-32-6), Tl (7440-28-0), V (7440-62-2), Zn (7440-66-6). Note: Hg is not typically analyzed by this method and is not recommended for low-level quantitative analysis; however, this method can be used as a screening tool (e.g., prior to analysis by a low-level method when high concentrations of mercury are expected). CAUTION: Also note that mercury memory effects may result from the analysis of samples that contain high level Hg concentration. See Method 6020B Sections 7.20.11, 7.22.3, and 11.1 from SW-846 Compendium for guidance when analyzing for Hg.

6.5.4 Description

The method comprises the following main steps:
1. Preliminary treatment of most samples is necessary because of the complexity and variability of sample matrices. Groundwater samples which have been pre-filtered and acidified will not need acid digestion. Samples which are not digested must either use an internal standard or be matrix-matched with the standards (i.e., acid concentrations should match).
2. Profile and calibrate the instrument according to the instrument manufacturer's recommended procedures, using the typical mixed-calibration standard solutions described in Section 7.11.1. Prepare the calibration curve as detailed in Section 10.7. Flush the system between each standard using the calibration blank (Section 7.11.2.1), or as the manufacturer recommends. In order to reduce random error, use the average intensity of multiple exposures for both standardization and sample analysis.
3. For all analytes and determinations, the laboratory must analyze an ICV (initial calibration verification, Sections 7.11.3 and 10.8.1) and a CCV (continuing calibration verification, Sections 7.11.4 and 10.8.5) and CCB (continuing calibration blank, Sections 7.11.2.1 and 10.8.5) after every 10 samples and at the end of the analysis batch run.
4. Analyze the samples and record the results. In between each sample or standard, rinse the system using the calibration blank solution (Section 7.11.2.1). Use a minimum rinse time of 1 minute. Each laboratory may establish a reduction in the rinse time following a suitable demonstration.

5. Determination of percent dry weight. When sample results are to be calculated on a dry-weight basis, a separate portion of sample for this determination should be weighed out at the same time as the portion used for analytical determination. CAUTION: The drying oven should be contained in a hood or vented. Significant laboratory contamination may result from a heavily contaminated hazardous waste sample. Immediately after weighing the sample aliquot to be digested, weigh an additional 5- to 10-g aliquot of the sample into a tared crucible. Dry this aliquot overnight at 105°C. Allow the sample to cool in a desiccator before weighing. Calculate the % dry weight as follows by Eq. (6.1):

$$\% \text{ dry weight} = \frac{\text{(g) of dry sample}}{\text{m(g) of sample}} \times 100 \qquad (6.1)$$

This oven-dried aliquot is not used for the extraction and should be appropriately disposed of once the dry weight is determined.

> **Critical evaluation:** *This is a very common and useful analytical method for inorganic determination in environmental analysis, for example, water. Physical interferences are effects associated with the sample nebulization and transport processes. Changes in viscosity and surface tension can cause significant inaccuracies, especially in samples containing high dissolved solids or acid concentrations. The large use of reactants—especially mineral acids—make it a method needed to be greenish, according to the green chemistry principles* (Chapter 5: Main analytical techniques).

Fig. 6.10 describes the simplified flowchart of the method.

6.6 Analytical method for inorganic residue in juice

Interest in the determination of inorganic arsenic (iAs) in products for human consumption such as food commodities, wine, and seaweed among others is fueled by the wide recognition by the World Health Organization and the Food and Agriculture Organization of the United Nations[3] of its toxic effects on humans, even at low concentrations

---
[3] Joint FAO/WHO (Food and Agriculture Organisation of the United Nations/World Health Organization) (2014). Food standards programme. Codex Alimentarius Commission. Report of the Eighth Session of the Codex Committee on Contaminants in Foods. Rep. 14/CF. Rome, Italy: FAO/WHO

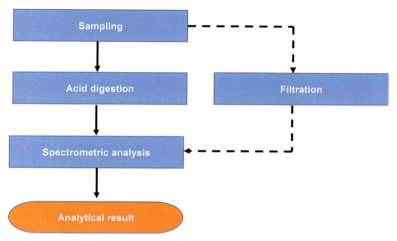

Figure 6.10 *Modus operandi* of the multielemental method for elements in the environment, especially in the water matrix (in some cases, analysis without previous digestion). *Credit: author.*

(Llorente-Mirandes, Rubio, & López-Sánchez, 2017). Other food groups that are important contributors to iAs exposure are rice, milk, and dairy products (the main contributor in infants and toddlers), and drinking water.

6.6.1 Method
Arsenic speciation using high performance liquid chromatography-inductively coupled plasma–mass spectrometric determination.

6.6.2 Source
U.S Food & Drug Administration (2020).

6.6.3 Scope and application
This method is used to learn how arsenic that may be present in juice is distributed among different chemical forms, or species; and, most importantly, how much is in the iAs forms. The method could be called for after an elevated level of total arsenic is found. Or, speciation information could be desired in routine monitoring. High performance liquid chromatography (HPLC) is used in combination with inductively coupled plasma–mass spectrometry (ICP-MS) to analyze fruit juice and determine mass fractions for arsenic species. The method targets two iAs species: arsenite, As(III), and arsenate, As(V); and two organic arsenic species—dimethylarsinic acid

(DMA) and monomethylarsonic acid (MMA). Typically, and depending on laboratory customer needs, the two inorganic species are summed and reported simply as total iAs. Should other (unknown) arsenic species be detected, it is also possible to estimate mass fractions for these. Juice is analyzed in ready-to-drink (RTD) condition. When received in concentrated form, the juice is reconstituted.

6.6.4 Description

RTD clear (i.e., no solids) juice is prepared by diluting an analytical portion approximately five-fold with water. Juice concentrates are diluted to a reconstituted RTD state prior to this five-fold dilution. Arsenic species are analyzed by HPLC-ICP-MS (Fig. 6.11) (Delafiori, Ring, & Furey, 2016). The HPLC uses a PRPX100 anion exchange column for separation. Arsenic species are identified by peak retention time match with arsenic species standards. The ICP-MS is used as an arsenic-specific detector to monitor for arsenic-containing chromatographic peaks. It is operated in helium collision cell mode or oxygen reaction mode, either of which minimize interference from co-eluting chloride species, which are commonly encountered in food analysis. Mass fractions are calculated based on peak area from analytical solutions compared to response of standard solutions. A (separate) total arsenic analytical method, is to be

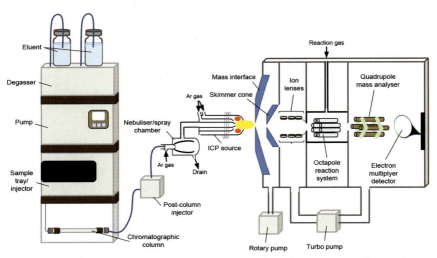

Figure 6.11 A HPLC-ICP-MS system. *Reproduced with permission from Elsevier, Delafiori, J., Ring, G., & Furey, A. (2016). Clinical applications of HPLC–ICP-MS element speciation: A review.* Talanta, 153, 306–331.

performed in conjunction with the species analysis—this method—so that when total arsenic is $\geq 10\ \mu g\ kg^{-1}$ (in RTD condition), the sum of the individual arsenic species mass fractions can be compared with the total arsenic mass fraction. This calculation indicates whether the arsenic is adequately accounted for.

The method comprises the following steps:

1. Sample preparation: (a.1) Dilute and shoot procedure—(i) Pipet 2 mL ($\sim$ 2 g) RTD juice into a tared 15-mL polypropylene centrifuge tube (record analytical portion mass), dilute to $\sim$ 10 g with DIW (record analytical solution mass), cap, and mix thoroughly. (ii) Draw at least 4 mL of the analytical solution into a syringe, attach a 0.45 μm Nylon or PTFE syringe filter, • discard the first $\sim$ 1 mL to waste • transfer $\sim$ 1 mL to an autosampler vial (for analysis) • dispense remainder (reserve) into a second 15 mL polypropylene centrifuge tube. iii. Store the reserve for up to 48 hours at 4°C in the event this sample solution needs to be re-analyzed. (a.2) Acid extraction procedure—(i) Weigh a 50 mL tube with lid and record the mass. (ii) Pipet 5 mL ($\sim$ 5 g) RTD juice into the pre-weighed 50 mL tube (record the mass). (iii) Add 6 mL 0.28 mol/L HNO_3, cap tube tightly, and mix thoroughly (vortex $\sim$ 15 seconds). (iv) Place tubes in hot block (preheated to 95°C) for 90 minutes. (v) Remove, cool, and add 9 mL DIW (record mass). (vi) Centrifuge (3000 rpm) for 10 minutes. g. Draw at least 4 mL of the analytical solution into a syringe, attach a 0.45 μm Nylon or PTFE syringe filter, and dispense into a second 15 mL polypropylene centrifuge tube (discard the first $\sim$ 1 mL to waste). (vii) Transfer $\sim$ 1 mL of the filtered extract to a tared 15 mL centrifuge tube (record mass), add $\sim$ 4 g DIW (record mass), and mix. (viii) Transfer $\sim$ 1 mL of diluted juice to an autosampler vial for analysis. (ix) Store the remainder for up to one week at 4°C in the event this sample solution needs to be re-analyzed. (Dilution factor is $\sim$ 20 following this procedure.)
2. Replicate analytical portions (RAPs): Use the sample preparation procedures described above for multiple (replicate) juice samples (dilute-and-shoot or acid extraction, as appropriate).
3. Fortified analytical portions (FAPs): Prepare an analytical portion fortified with As(III), DMA, MMA and As(V) at a level of 25 μg kg^{-1} each by combining 2 mL ($\sim$ 2 g) RTD juice and 0.05 mL ($\sim$ 0.05 g) of the 1000 ng g^{-1} multi-analyte spiking solution in a 15 mL polypropylene centrifuge tube. It is recommended that the same sample be

used for both FAP recovery and replicate precision. Proceed with analysis as would for unfortified sample (dilute-and-shoot or acid extraction, as appropriate).
4. Method blank (MBK): Take 2 g DIW through the sample preparation procedures described above for juice samples (dilute-and-shoot or acid extraction, as appropriate).
5. CRM: Depending on CRM matrix, prepare as if a sample but dilute (with DIW) appropriately according to the knownlevel so the analytical solution's level will be within the calibration. Although NIST SRM 1643f is not certified for arsenic species, As(V) and possibly As (III) should be the only peaks detected. For this CRM, a dilution factor of ~15 × is appropriate to dilute the acid content.
6. Analysis: Proceed according to the batch sequence. Standardize then inject/analyze the various QC solutions (MBKs, resolution check solution, CCV, RAPs, FAPs, CRMs) and sample solutions.
7. QC analyses: (1) Standardize using a minimum of four calibration levels for each arsenic species. (2) Analyze method blanks at start and with every 10 or fewer analytical solutions. (3) Analyze the CCV standard after every 10th analytical solution and at the end (after the last analytical solution analyzed) to monitor retention time and calibration. (4) Analyze two RAPs at a frequency of 10% (i.e., at least one RAP for every ten samples) and at least one for each type of juice. Analysis of a greater number of RAPs (e.g., triplicate) is permitted but not required. (5) Analyze at least one FAP for each type of juice to verify peak identification and quantitative recovery. (6) Include at least one CRM or in-house reference material in each batch. Since juice is comprised largely of water, reference materials such as NIST1643f trace elements in water[4] represent a reasonable matrix match.

Critical evaluation: *It is a very powerful speciation method for inorganic contaminants in food and related products—it can be adapted to several matrices. Its limitation is mainly due to the higher-cost. Additionally, as for ICP-OES it is a method to be greenish, according to the green chemistry principles (Chapter 5: Main analytical techniques).*

Fig. 6.12 illustrates the simplified flowchart of the method.

[4] https://www-s.nist.gov/srmors/view_detail.cfm?srm = 1643F

Analytical methods to selected matrices 191

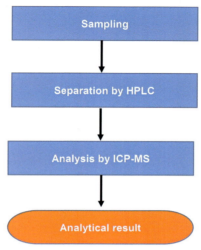

Figure 6.12 *Modus operandi* of the spectrometric method for arsenic speciation in food. *Credit: author.*

6.7 Analytical method for inorganic residue in meat

Regarding the occurrence of toxic metals in meat, the risk associated to cadmium (Cd) in organic production has been reported as the consequence of untreated raw phosphates (Guéguen & Pascal, 2010).

6.7.1 Method

Determination of cadmium and other 30 elements in foodstuffs by ICP-MS after closed-vessel microwave digestion.

6.7.2 Source

Chevallier, Chekri, Zinck, Guérin, and Noël (2015).

6.7.3 Scope and application

Cadmium analysis in foodstuffs with focus on meat as the matrix of interest.

6.7.4 Description

The method is composed by the following steps:
1. Samples are digested using a microwave digestion system equipped with a rotor for eight 80 mL quartz vessels (operating pressure, 80 bar).

Figure 6.13 An ICP-MS equipment (*left*) with a microwave digester device (*right*). Courtesy Shimadzu.

The sample digestion procedure is performed according to the NF EN 13805 standard.[5] Before use, the quartz vessels are decontaminated with 6 mL of 50% HNO_3 (54%, v/v) in the microwave digestion system, then rinsed with ultrapure water and dried in a 40°C oven. From 0.3 to 0.5 g of meat samples should be weighed precisely in the quartz digestion vessels and wet-oxidized with 3 mL ultrapure water and 3 mL suprapur HNO_3 (67%) in the microwave digestion system. One randomly selected vessel is filled with reactants only and taken through the entire procedure as a blank. After cooling to room temperature, sample solutions are transferred into 50 mL polyethylene flasks. Then, 100 mL of the internal standard solution (1 mg L^{-1} internal standards and 10 mg L^{-1} Au) are added to a final concentration of 2 mg L^{-1} internal standards and 20 mg L^{-1} Au; the digested samples are filled with ultrapure water to the final volume before analysis by ICP-MS.

2. ICP-MS measurements are performed with a spectrometer equipped with a reaction system (preferably) using He gas (Fig. 6.13). The sample solutions are pumped by a peristaltic pump from tubes arranged on an auto-sampler. The torch position, ion lenses, gas output, resolution axis (10% of peak height, m $\pm$ 0.05 a.m.u) and background (<20 shots) are optimized daily with the tuning solution (1 mg L^{-1}) to carry out a short-term stability test of the instrument, to maximize ion signals and to minimize interference effects due to high oxide levels (CeO^+/Ce^+ < 1.2%) and doubly charged ions (Ce^{2+}/Ce^+ < 2%).

---

[5] https://www.en-standard.eu/din-en-13805-foodstuffs-determination-of-trace-elements-pressure-digestion/?gclid = EAIaIQobChMIro6Ju5nI7AIVFAeRCh0uZg38EAMYASAA EgL0dPD_BwE

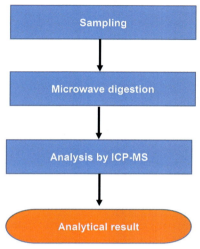

Figure 6.14 *Modus operandi* of the spectrometric method for cadmium in meat. *Credit: author.*

Linearity response in the pulsed and analog modes (P/A factor determination) is verified daily using tuning solutions.

Critical evaluation: *As a typical ICP-MS method, it can provide low LOD and LOQ for the analyte. Acid digestion requires care in its handling. However microwave digestion can be seen as a green technique that reduces energy and residue generation.*

Fig. 6.14 illustrates the simplified flowchart for the method.

6.8 Conclusion

Analytical methods based on instrumental techniques, especially chromatography and spectrometry, are very useful in the analysis of organic and inorganic residues from agrochemicals in food and environment-related matrices.

These methods can be adapted to several matrices according to the analyst's understanding and their technical expertise which delivers a versatile set of tools for quality control and monitoring. On the other hand, it is highly desirable that such methods follow the principles of green chemistry.

References

Chevallier, E., Chekri, R., Zinck, J., Guérin, T., & Noël, L. (2015). Simultaneous determination of 31 elements in foodstuffs by ICP-MS after closed-vessel microwave digestion: Method validation based on the accuracy profile. *Journal of Food Composition and Analysis, 41*, 35−41.

Delafiori, J., Ring, G., & Furey, A. (2016). Clinical applications of HPLC−ICP-MS element speciation: A review. *Talanta, 153*, 306−331.

European Union. (2020). *Feed additives*. <https://ec.europa.eu/food/safety/animal-feed/feed-additives_en> Accessed 10.20.

Ferrer, I., & Thurman, E. M. (2007). Multi-residue method for the analysis of 101 pesticides and their degradates in food and water samples by liquid chromatography/time-of-flight mass spectrometry. *Journal of Chromatography A, 1175*, 24−37.

Guéguen, L., & Pascal, G. (2010). An update on the nutritional and health value of organic foods. *Cahiers de Nutrition et de Diététique, 45*, 130−143.

International Atomic Energy Agency. (2016). *Manual of standard operating procedures for veterinary drug residue analysis*. Vienna.: IAEA.

Llorente-Mirandes, T., Rubio, R., & López-Sánchez. (2017). Inorganic arsenic determination in food: A review of analytical proposals and quality assessment over the last six years. *Applied Spectroscopy, 71*, 25−69.

Tikhomirova, T. I., Ramazanova, G. R., & Apyari, V. V. (2017). A hybrid sorption−spectrometric method for determination of synthetic anionic dyes in foodstuffs. *Food Chemistry, 221*, 351−355.

U.S. Environmental Protection Agency. (2007). *Method 8081B—organochlorine pesticides by gas chromatography*. <https://www.epa.gov/sites/production/files/2015-12/documents/8081b.pdf> Accessed 10.20.

U.S Food & Drug Administration. (2020). *Elemental analysis manual for food and related products. 4.10 High performance liquid chromatography inductively coupled plasma-mass spectrometric determination of arsenic species in fruit juice*. <https://www.fda.gov/food/laboratory-methods-food/elemental-analysis-manual-eam-food-and-related-products> Accessed 10.20.

CHAPTER 7

Analytical chemistry towards a sustainable agrochemistry

Nowadays, agriculture is constantly required to become more sustainable, with reduction of its negative impacts on the environment and public health allied to increasing their positive impacts on society and the economy. These are challenges and, at the same time, opportunities for new production systems for those agrochemicals are indispensable.

How could it not be different, these opportunities are extended to the use of emerging technologies and approaches to support a sustainable agrochemistry proposal in order to reduce negative impacts from agrochemicals to public health (e.g., occupational diseases) and to the environment (e.g., pollution of air, soil, and water).

7.1 What is sustainable agrochemistry?

The practice of agriculture is one of the oldest activities developed by humans. In the Neolithic Period, the constitution of the first techniques and materials used for the cultivation of plants and the confinement of animals was the main cause for what was denominated as the sedentarization of humans. It allowed fixing residence in a given locality, although collection and hunting have long coexisted side by side with agriculture.

The development of agriculture, therefore, was directly associated with the formation of the first civilizations, which helps us understand the importance of techniques and the environment in the process of building societies and their geographical spaces. In that sense, as these societies further advanced their techniques and technologies, the more the evolution of agriculture was benefited.

Originally, the practice of farming was developed in the vicinity of large rivers, notably the Tigris and Euphrates, as well as the Nile, the Ganges, and others. Not coincidentally, it was in these localities that the first great known civilizations emerged, because the practice of agriculture allowed the development of trade to a surplus in production.

One of the most important moments in the process of agricultural evolution throughout history was, without doubt, what became known as the Agricultural Revolution (British Broadcasting Corporation, 2017). We can say that, over time, several agricultural revolutions have succeeded, but the main ones occurred after the Industrial Revolution.

The process of industrialization of societies allowed the transformation of the geographic space in the rural environment, which occurred thanks to the insertion of greater technological apparatuses in the agricultural production, allowing a greater mechanization of the field. This transformation materialized from the supply of inputs from the industry to agriculture, such as machinery, fertilizers and technical objects and practices in general.

The influence of the Agricultural Revolution in the world was also directly associated with the European maritime-colonial expansion, in which the European peoples disseminated their different cultures through the world by means of introducing new crops and novel agricultural practices. It is worth remembering that the interaction between settlers and colonizers also contributed to the agricultural evolution, in as much as previously little known techniques were applied and disseminated, such as the terracing practiced in both ancient China and pre-Columbian civilizations.

As an effect of this revolution, chemistry started to have a fundamental role in the agriculture expansion and technological development by means of fertilizers, pest controllers, scientific knowledge, among other aspects related to production systems.

In the 20th century, more precisely after World War II, the evolution of agriculture reached one of its most important hallmarks, in what became known as the Green Revolution. As observed by Pingali (2012), it was based on a set of measures and promotion of techniques based on the introduction of genetic improvements in plants and the evolution of agricultural production apparatus to expand, above all, food production.

The introduction of techniques from the Green Revolution led to a large-scale increase in grain and cereal production, significantly reducing the need for food in various regions of Asia, Africa, and Latin America, even though hunger has not been eradicated, since its existence is not only due to lack of food. The impact on the world was so wide that the American agronomist Norman Borlaug, considered the "father" of the Green Revolution, was awarded the Nobel Peace Prize in 1990.

Although the Green Revolution is heavily criticized for its environmental impacts and the process of land concentration that accompanied its evolution due to policies that were used to promote rapid intensification of agricultural systems and increase food supplies (Pingali, 2012), its importance for the development of agriculture in the world is undeniable. Furthermore, in the following decades, the improvements resulting from technology in the field, such as biotechnology and mechanization, have increased in the following decades, which has been further increasing the productivity.

7.1.1 The demand for sustainability

Nowadays, agriculture is constantly required to become more sustainable, with reduction of its negative impacts on environment allied to an increasing in their positive impacts on society and economy. These are challenges and, at same time, opportunities for new production systems.

For a conceptual understanding, we can define *agrochemistry* as the application of chemistry in agriculture. Its action, object of study and technical application is not only concerned with the production of agrochemicals but also with the analysis and prevention of the harmful effects of chemical substances on both crops and humans (farmers and consumers). A definition of this concept can be seen in the Fig. 7.1 with contribution and intersection with food & agrosciences, biotechnology, nanotechnology, chemical sciences and—of course—sustainability pillars comprising economic, environmental and social impacts.

Analytical & environmental chemistry allied to technological chemistry input techniques, technologies and knowledge to analyze, produce and monitoring agrochemicals. Nanotechnology and biotechnology are new technological approaches to be incorporated to the agrochemicals for the best agronomic usages. Food security and environment are closely related to the agrochemical effects and near to the consumer perception, implying in laws, market restrictions and public opinion. Finally, sustainability is a demand of the society for greater quality of life and greater transparency in the productive chains.

Historically, the contribution of chemistry goes back to the 19th century, with the synthesis of inorganic fertilizers and, by the middle of the last century, of a great number of compounds synthesized to control insects, diseases, and weeds (Pinto-Zaervallos & Zarbin, 2013).

This contribution can be seen clearly and decisively in the cycle of nitrogen, an essential element to some molecules that integrate the organic matter. Plants, with some exceptions, do not have the capacity to

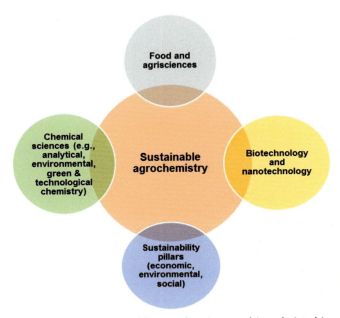

Figure 7.1 The concept of sustainable agrochemistry and its relationship with areas of technological and scientific knowledge. *Credit: author.*

absorb this element from the atmosphere (with 78% v/v of nitrogen), the opposite of what occurs with another essential element, carbon, which is absorbed as CO_2 by means the photosynthesis. So, the only natural way to close the nitrogen cycle is through the decomposition of organic material from dead animals or plants, or the excretion of living organisms. However, this form of replenishment is naturally limited.

The capture and use of atmospheric nitrogen in the soil was only possible economically via the works of the German chemists Fritz Haber and Carl Bosch. They developed the Haber-Bosch reaction or process at the beginning of the 20th century (Ritter, 2008), which allows the ammonia synthesis from the small reactive atmospheric nitrogen and another abundant element, hydrogen, in industrial scale. Curiously, the incentive that led to this essential innovation was not initially the production of fertilizers, but the production of nitrates for military purposes (i.e., explosives) to be used in World War I.

Ammonia, a molecule with about 82% by weight of nitrogen, can be absorbed by plants through the soil. Although, for reasons of ease of application, it is preferred to be used as nitrogenous fertilizers solid substances derived from ammonia, such as urea and ammonium nitrate. The world production of ammonia today reaches 150 million tons per year (U.S. Geological Survey, 2020), and almost all global production is destined to the synthesis of industrial fertilizers. The percentage of the world population whose food depends on the use of synthetic nitrogen fertilizers is estimated at 53% (Liu, Ma, Ciais, & Polasky, 2016), generating a high demand for these agricultural inputs.

Taking the nitrogen fertilizers as an example, we can also conclude that so-called organic farming, which advocates the exclusion of synthetic fertilizers, can function as a niche market in societies of abundance, but it is certainly not an alternative to feed humanity. Then, we can see clearly the contribution of the agrochemistry to the well-being of modern society. A sustainable agrochemistry can represent a holistic view for a modern agricultural practice.

7.1.2 The agrochemical classes

For a better knowledge of the agrochemical demand and the necessity of a more sustainable agricultural systems, is useful considering its definition. According the International Union of Pure and Applied Chemistry (2006), an agrochemical is an *"agricultural chemical used in crop and food production including pesticide, feed additive, chemical fertilizer, veterinary drug, and related compounds"*.

Currently, we can observe several agrochemical classes according their uses in agriculture:
- Fertilizers—any kind of substance applied to soil or plant tissues to provide one or more nutrients essential to plant growth;
- Plant growth regulators—(also called plant hormones)—several chemical substances that profoundly influence the growth and differentiation of plant cells, tissues, and organs;
- Phytosanitary products, pesticides or correctives:
  * Herbicides—agents, usually chemicals, used for killing or inhibiting the growth of unwanted plants (i.e., weeds);
  * Insecticides—pesticides formulated to kill, harm, repel or mitigate one or more species of insects;
  * Fungicides—pesticides that kill or prevent the growth of fungi and their spores;

- Acaricides—pesticides that kill members of the arachnid subclass Acari, which includes ticks and mites;
- Bactericides—a substance that kills bacteria. *Bactericides* are disinfectants, antiseptics, or antibiotics;
- Rodenticides—pesticides that kill rodents. Rodents include not only rats and mice, but also squirrels, woodchucks, chipmunks, porcupines, etc.;
- Nematicides—a type of chemical pesticide used to kill plant-parasitic nematodes;
- Repellents—chemicals that can help reduce the risk of being bitten by insects and therefore reduce the risk of getting a disease carried by mosquitos or ticks;
- Fumigants—any volatile, poisonous substance used to kill insects, nematodes, and other animals or plants that damage stored foods or seeds;
- Disinfectants—antimicrobial agents that are applied to the surface of non-living objects to destroy microorganisms that are living on such objects;
- Antibiotics—powerful drugs that fight bacterial infections; as a highlighted risk, their continued use may result in resistance to them by some microorganisms;
- Defoliants—a chemical dust or spray applied to plants to cause their leaves to drop off prematurely;
- Algaecides—or algicide—a biocide used for killing and preventing the growth of algae.

Agrochemicals move a huge global market. According to Statista (2020), the global agrochemical market for crop protection in 2019 was worth some 59.8 billion U.S. dollars. However, they are one of the main classes of chemical pollutants, with serious negative impacts on public health and the environment. The search for alternatives to conventional agrochemicals presents itself as an excellent opportunity for the development of sustainable agricultural technologies and for opening new businesses.

Governments, farmers, and consumers show increasing concerns worldwide with the negative impacts on the environment and health caused by the large amount of inputs applied to produce different crops. Agrochemicals have a direct correlation with damages from agriculture, with pesticides being the main representative class with toxicological implications.

7.1.3 Undesirable effects from agrochemical usages

In order to visualize the extension of deleterious effects from agrochemicals, it is necessary to have a definition of negative impacts. Firstly, the main negative impacts from the agriculture on the environment are:
- Water and air pollution due to the use of pesticides;
- Extinction of water bodies, due to high water demand;
- Erosion and soil degradation due to inadequate management during cultivation;
- Change in biota due to factors already listed;
- Changes in the quality of environmental resources, also due to factors already listed;
- Ecological risks for insects, plants and animals associated with the change of environment;
- Climate change due to the deforestation and biomass combustion.

Regarding those impacts on health, it can be highlighted:
- Poisoning due to pesticide use and food contaminated consumption;
- Occupational risks to farmers due to the exposition to pesticides;
- Human infections—or emerging infectious diseases -, that do not respond to treatment due to the use of antimicrobials in agriculture (Grace, 2019).

From these negative impacts, it is becoming paramount the development of a more environmentally and healthy friendly agriculture.

Agricultural chemistry is, undoubtedly, one of the fields of research and business whose impact is felt throughout the world, since we all need to eat to survive. Added to this is the fact that, increasingly, technology is intertwining with modern agriculture, both with regard to new production strategies and the reduction of negative environmental impacts (Herman, 2015).

7.1.4 Pillars of sustainability

Sustainability can be seen and understood by means of its three components or pillars:
- Environmental impacts
- Economic impacts
- Societal impacts

Impacts can be positive or negative according to their direct or indirect effects upon the environment, economy, and society. Considering that agrochemistry is the application of chemistry and its concepts and technologies to promote a better agriculture, economic, and societal impacts are expected to be positive, especially the economic impacts. On the other hand, and due to a historic of incidents at global level, environmental impacts are expected to be negative; nevertheless, it could be positive if modern technologies and good agricultural practices are used. A more detailed evaluation of sustainability in agriculture can be seen in Quintero-Angel and González-Acevedo (2018).

Sustainable chemistry, a recent branch of chemistry, was defined as *"(...) a scientific concept that seeks to improve the efficiency with which natural resources are used to meet human needs for chemical products and services. Sustainable chemistry encompasses the design, manufacture and use of efficient, effective, safe and more environmentally benign chemical products and processes."* (Organization of Economic Co-operation and Development, 2021a, 2021b). From these statements, a relationship with agrochemistry can be constructed by means the design, manufacture and use of efficient, effective, safe and more environmentally benign agrochemicals (Vaz Jr., 2019). That is, by the establishment of a strong innovation drive in agriculture for the next decades.

7.1.5 Agrochemical regulation and commercialization

Agrochemical regulation and commercialization is very desirable for the reduction of the negative impacts to reach a sustainable condition. Table 7.1 describes some commercial agrochemical formulations and their environmental classification of dangerousness that emphasizes the need of new formulations and/or application strategies, in accordance with new trends as circular economy to make sustainable products, focusing on the sectors that use the most resources and where the potential for circularity is high such as food, water, and nutrients (European Commission, 2021).

We can observe the presence of "other ingredients" term in the chemical composition which turns the agrochemical formulation a "black box".

Generally, a pesticide formulation comprises:
- Active ingredient (AI): Substance that gives effectiveness to the formulated product;
- Adjuvants: Substances that improve the performance of formulated products, for example, surfactant, compatibility agent, buffer, promoting the AI reach its maximum effectiveness;
- Inert: Non-reactive substances with other components of the mixture, for example, water.

Table 7.1 Chemical composition and the environmental dangerousness of some commercial agrochemicals.

Product	Active ingredient (AI)	Use	Composition[a]	Environmental classification[a]
Heat (BASF)	Saflufenacil	Herbicide	AI + other ingredients	Very dangerous product
Curbix (BAYER)	Etiprole	Insecticide	AI + other ingredients	Very dangerous product
Aumenax (BASF)	Fluxapiroxade + copper oxychloride	Fungicide	AI + other ingredients	Very dangerous product

[a] According to the product label from the producer, available on: www.agrofit.agricultura.gov.br.

In some cases, it is necessary analyze and monitoring not only the presence and effect of the AI, but also the adjuvant because the last can be more harmful than the first. That means, the necessity of new formulations involves the AI and other ingredients. For instance, Mesnage and Antoniou (2018) observed that despite the known toxicity of adjuvants, they are regulated differently from AIs, with their toxic effects being generally ignored leading to a misrepresentation of the safety profile of commercial pesticides. Moreover, adjuvants are not subject to an acceptable daily intake, and they are not included in the health risk assessment of dietary exposures to pesticide residues.

All agrochemicals, especially pesticides, need to be registered for their commercialization with each country having its own agencies and rules for that—a national pesticide legislation. However, FAO[1] and OECD[2] have general proceedings to guide this registration taking into account the aspects of ecotoxicology, toxicology, processing, and quality control.

According FAO, the principles of pesticide registration include (United Nations Food and Agriculture Organization, 2010):

- Comprehensive, harmonized, and clear registration requirements and criteria;
- Use of all available information and mutual acceptance of data;
- Transparency and exchange of information;
- Science-based assessment to determine whether precautionary approaches are warranted;
- Consideration of hazard;
- Risk assessment and mitigation based on the local situation;
- Risk—benefit analysis, taking into consideration the availability of alternatives;
- Postregistration monitoring and evaluation;
- Mechanisms for periodic and unscheduled review in order to respond to new information that may affect the regulatory status;
- Protection of intellectual property rights of the application;
- Evaluation of the data submitted in the registration dossier should follow internationally accepted and agreed evaluation standards and procedures.

It is important to highlight the necessity of science-based assessment in these principles in order to outstand the evidence of the use of analytical

---

[1] http://www.fao.org/pesticide-registration-toolkit/en/
[2] https://www.oecd.org/env/ehs/pesticides-biocides/agriculturalpesticidesprogramme.htm

technologies to contribute to the sustainability for agriculture and related sectors.

Regarding to tests and analyses, they will follow the good laboratory practices in order to reach the required reliability of submitted data (Organization of Economic Co-operation and Development, 2021a, 2021b).

For the determination of **residues in rotational crops**, Fig. 7.2 illustrates the flowchart strategy—a sorghum crop is a good example of a rotational crop (Fig. 7.3). This strategy is divided in three tiers, where Tier 1 and Tier 2 have a direct necessity of the analytical data generating using, for instance, chromatographic techniques to conduct it.

Nowadays, a risk to sustainable agrochemicals are the commercialization and use of stolen or smuggled pesticides. It is very common in developing countries with a high agricultural production—that means, a high demand for agrochemicals—allied to an extensive area with large boarders without suited vigilance or no control over the production of agrochemicals, as Latin America.

The trade of these products causes a lack of control over the chemical compounds used in crops, often leading citizens to consume food with prohibited substances and, obviously, harmful to health. It is clear that an absence in the control of fabrication and commercialization will generate more negative impacts on the environment and society, reducing or eliminating the sustainability condition. However, the presence or absence of illegal pesticides don't promote the lack of toxicity/ecotoxicity associated to the legal pesticides—the illegals just will increase the effects.

As seen in this introductory item, analytical chemistry is mandatory to guarantee a condition of sustainability in agriculture because it is the only scientific way to supply information in atomic and molecular levels to promoting monitoring, control, and development of chemical compounds.

7.2 Outlook of emerging analytical technologies

Emerging analytical technologies or instrumentation can be understood as a set of analytical techniques which are not usual in routine analysis laboratories but have a potential to reach out in a near future.

They can be comprised in those common classes as described in Chapter 5, Main Analytical Techniques, for analytical instrumentation, highlighting spectroscopic and electrochemical fundamentals of operation. However, they are closely related to the cutting-edge analytical technology. On this context, *nonosensors* are especially of interest due to a promise of low cost allied to rapidness and easy-to-handle characteristics.

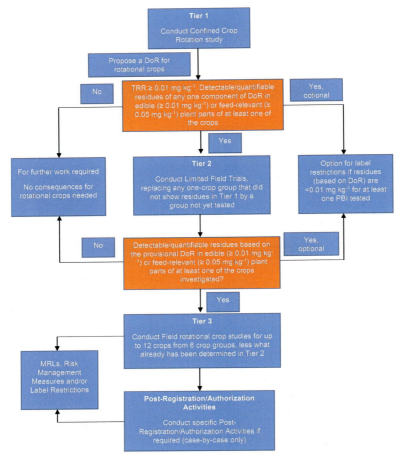

Figure 7.2 The general decision tree (DT) based on the tiered approach for residues in rotational crops. DoR, definition of residue; MRLs, maximum residue limits. *Reproduced with permission from Organisation for Economic Co-operation and Development. (2018).* Series on pesticides no. 97 and series on testing and assessment no. 279 Guidance document on residues in rotational crops. *Paris: OECD.*

According Kumar, Guleria, and Mehta (2017), nanosensors can detect:
- Pathogenic bacteria
- Food-contaminating toxins
- Adulterant
- Vitamins
- Dyes
- Fertilizers

Figure 7.3 A mechanical herbicide application in a sorghum crop (*Sorghum bicolor* [L.] Moench) in the Brazilian tropical region of mid-west. *Credit: author.*

- Pesticides
- Taste and smell

As the book subject is the analysis of chemical residues from agrochemicals in food and in the environment, the interest are those nanosensors able to detect fertilizers and pesticide residues, especially the last ones. Table 7.2 describes examples of these applications.

In order to found the better analytical technology, Munawar et al. (2019) highlight the unique and exceptional properties of nanomaterials (large surface area to volume ratio, composition, charge, reactive sites, physical structure and potential) to be exploited for sensing purposes. High-sensitivity in analyte recognition is achieved by preprocessing of samples, signal amplification and by applying different transduction approaches.

Fig. 7.4 depicts a simplified flowchart for nanosensor application in the agrochemical residues detection.

Despite the facility of application of nanosensors, they cannot substitute the laboratorial techniques (e.g., chromatography) for a complete investigation, as showed in the item 7.3.

7.2.1 Artificial intelligence in analytical chemistry

Artificial intelligence (AI) is an emerging technology applied to data planning and treatment. As an example of AI, chemometrics was detailed in

Table 7.2 Examples of nanosensors use for detection of agrochemical and food additive residues.

Type of nanoparticle	Technique	Analyte
MWCNT, graphene and iron oxide	Cyclic voltammetry and solid-phase extraction—high-performance liquid chromatography	Sudan I
Liposome, gold, zirconium dioxide—gold and zinc sulfide—cadmium selenide and thioglycolicacid—cadmium selenide	Square wave voltammetry; colorimetry, fluorescence, photoluminescence and ultraviolet—visible spectroscopy	Organophosphorus: parathion, paraoxon, and carbamate pesticides
Silver and gold	Colorimetry, fluorescence and ultraviolet—visible spectroscopy	Melamine
Cobalt nitroprusside	Cyclic voltammetry	Sulfite
SWCNT, MWCNT—silica, MWCNT—zinc oxide, MWCNT—platinum and MWCNT—ionic liquids	Cyclic voltammetry and field-effect transistor	Bisphenol A, cadmium ions, sunset yellow, and tartrazine
Poly(ethylene glycol dimethacrylate-*N*-methacryloyl-L-histidine methylester)	Ultraviolet—visible spectroscopy; surface plasmon resonance	Chloramphenicol

MWCNT, multi wall carbon nanotubes; *SWCNT*, single wall carbon nanotubes.
Adapted from Kumar, V., Guleria, P., & Mehta, S. K. (2017). Nanosensors for food quality and safety assessment. *Environmental Chemistry Letters*, 15, 165—177. Reprinted with permission from Springer Nature.

the Chapter 4, Fundamentals of Analytical Chemistry. However, AI is more than chemometrics as we will see here.

In a simple way, AI can be understood as the use of mathematical and computational tools—for example, machine learning—to predict behaviors from measuring data. And it can be a useful set of techniques to obtain the best information from a certain analytical process for to a certain problem solving (e.g., quality control of soil or food). Then AI can contribute to a construction of smart analytical laboratories and can avoid human error generation during sample preparation and analysis by means algorithms of

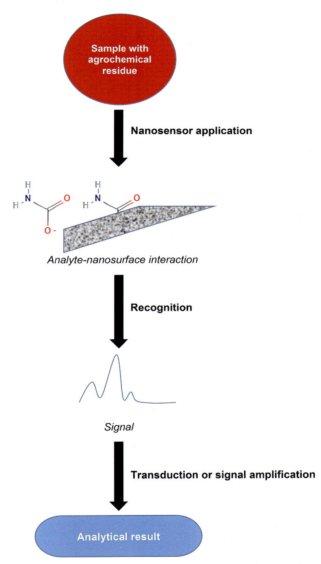

Figure 7.4 Simplified flowchart of a nanosensor application for agrochemical residue analysis. *Credit: author.*

automatization, what are fundaments followed by an analytical chemist, as seen in Chapter 4.

In fact, AI is not recent and Hippe (1983) pointed out some problems to be considered in its application in analytical chemistry dedicated to organic analysis: "*Particular attention is given to the approaches based on problem-reduction search methods, algorithmic and heuristic, in elucidation of the structure of complex organic compounds by sophisticated computer program systems.*"

As a practical application, Dotzert (2020) states that AI can support chromatographic method development. It encompasses the selection of a chromatography mode, detector, stationary phase, mobile phase, and numerous other factors. Method development becoming a more frequent task in analytical laboratories, as new target molecules continue to emerge.

Vasseghian, Berkani, Almomani, and Dragoi (2021) applied AI for data mining to generate optimum predictive models for pesticide decontamination processes—a paramount strategy to reduce the negative impact on the environment—using heterogeneous photocatalytic processes. In the present study, 537 valid cases from 45 articles from January 2000 to April 2020 were filtered based on their content collected and analyzed. Based on cross-industry standard process (CRISP) methodology, a set of four classifiers were applied: decision trees (DT), bayesian network (BN), support vector machines (SVM), and feed forward multilayer perceptron neural networks (MLP). To compare the accuracy of the selected algorithms, accuracy, and sensitivity criteria were applied. After the final analysis, the DT classification algorithm with seven factors of prediction, the accuracy of 91.06%, and sensitivity of 80.32% was selected as the optimal predictor model—an example of DT is presented in the Fig. 7.2.

7.3 Outlook of innovative analytical approaches

Automation and miniaturization of analytical techniques—for example, chromatographic and spectroscopic techniques—are technological approaches always in development and evidence. Automation gives the opportunity to reduce, for instance, LOD and LOQ, time and costs—despite sought, they are not always achieved—while miniaturization gives the opportunity to reduce the consumption of reactants and solvents.

These strategies are very applicable to monitoring (miniaturization) and to quality control (automation). Hyphenation—for example, GCxGC-MS, SPME-GC-FID, LC-MS^n—is a typical example for automation while sensors and probes are for miniaturization. Both approaches can be combined to reach the best analytical condition and results and are subject for application of green chemistry principles (American Chemical Society, 2021), in order to reduce negative environmental impacts from analytical processes. These approaches will guide this item.

7.3.1 Combined automation and miniaturization approach for surfactant presence

4-nonylphenol is an estrogenic endocrine active chemical that is a derivative from ethoxylated alkylphenol, a non-ionic surfactant present in pesticide formulation as adjuvant and is known to contaminate food and drinking water. It's considered also an indirect additive used in food contact substances.

According Li, Jin, and Snyder (2018), the analysis of nonylphenols (NPs) and their precursors NP ethoxylates (NPEOs) (Fig. 7.5) in food and environment is an increasing concern from the viewpoint of risk assessment due to their endocrine-disrupting effects.

The miniaturized and automated sample preparation techniques, as exemplified by magnetic solid-phase extraction (SPE) and liquid phase microextraction (SPME) have become the promising technologies for analysis of organic residues (Fig. 7.6). Gas and liquid chromatography coupled to mass spectrometers are still predominant technologies for analysis of the targets, but the supercritical fluid chromatography coupled to tandem mass spectrometer that has emerged recently may become an alternative method. Authors highlighted the need for in-depth study of the isomer-specific toxicity, with the individual analysis of NPEO homologs and their degradation product NP isomers.

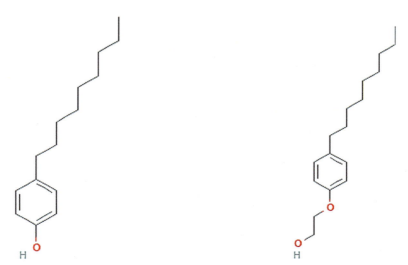

Figure 7.5 Chemical structures of nonylphenol (*left*) and its precursor nonylphenol ethoxylate (*right*). *Credit: author.*

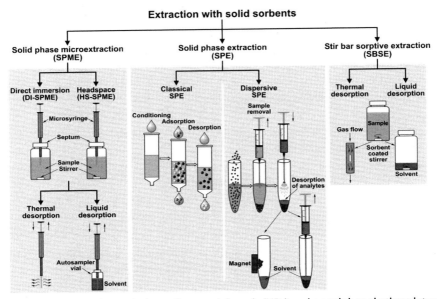

Figure 7.6 Extraction techniques for nonylphenols (NPs) and nonylphenol ethoxylates (NPEOs) using solid sorbents for chromatographic analyses. *Adapted from Grześkowiak, T., Czarczyńska-Goślińska, B., & Zgoła-Grześkowiak, A. (2016). Current approaches in sample preparation for trace analysis of selected endocrine-disrupting compounds: Focus on polychlorinated biphenyls, alkylphenols, and parabens. TrAC, Trends in Analytical Chemistry, 75, 209–226. Reproduced with permission from Elsevier.*

7.3.2 Combined automation and miniaturization approach for multiresidue presence

Bernardi, Kemmerich, Adaime, Prestes, and Zanella (2020) developed and validated a miniaturized sample preparation method for the multiresidue determination of 97 pesticides in wine samples. The proposed extraction procedure is based on the QuEChERS acetate method with dispersive solid phase extraction (d-SPE) for the clean-up step. Ultra-high performance liquid chromatography coupled with tandem mass spectrometry (UHPLC-MS/MS) was used for determination. The extraction and clean-up steps were evaluated to obtain the best conditions for the selected pesticides. Miniaturization of the sample preparation step provided a reduction in the consumption of samples and chemicals. The LOQ of the method was between 10 and 20 $\mu g\ L^{-1}$. Trueness results, obtained by recovery assays at the spike levels 10, 20, 50, and 100 $\mu g\ L^{-1}$, ranged from 70% to 120% with precision in terms of relative standard deviations (RSD) ≤ 20%. The method was successfully applied for the

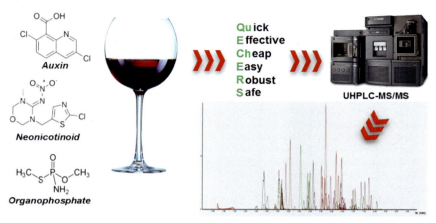

Figure 7.7 Flowchart for the use of miniaturized QuEChERS method for pesticide residue analysis in wine. *Adapted from Bernardi, G., Kemmerich, M., Adaime, M. B., Prestes, O. D., & Zanella, R. (2020). Miniaturized QuEChERS method for determination of 97 pesticide residues in wine by ultra-high performance liquid chromatography coupled with tandem mass spectrometry. Analytical Methods, 12, 2682–2692. Reprinted with permission from Royal Society of Chemistry.*

analysis of wine samples and different pesticides were found at concentrations from 14 to 55 µg L^{-1}.

Fig. 7.7 illustrates this approach.

7.4 Conclusion

Sustainability is, undoubtedly, mandatory for modern agriculture. Especially when considering agrochemical inputs, a source of potential pollutants for the environment and contaminants for food, feed, and beverages. And residue presence is a risk to be studied and evaluated by means of a holistic vision in order to guarantee its reduction or the desirable absence.

In order to reach the sustainability condition, emerging analytical technologies (e.g., nanosensors) can be used allied to automation and miniaturization approaches. Furthermore, AI is a set of tools to be exploited to supply more analytical information.

References

American Chemical Society. (2021). *Design principles*. <https://www.acs.org/content/acs/en/greenchemistry/principles.html> Accessed 1.21.

Bernardi, G., Kemmerich, M., Adaime, M. B., Prestes, O. D., & Zanella, R. (2020). Miniaturized QuEChERS method for determination of 97 pesticide residues in wine by ultra-high performance liquid chromatography coupled with tandem mass spectrometry. *Analytical Methods, 12*, 2682−2692.

British Broadcasting Corporation. (2017). *Agricultural revolution in England 1500−1850.* <http://www.bbc.co.uk/history/british/empire_seapower/agricultural_revolution_01.shtml> Accessed 1.21.

Dotzert, M. (2020). *The power of algorithms in analytical chemistry.* <https://www.labmanager.com/insights/the-power-of-algorithms-in-analytical-chemistry-23167#:~:text = AI%20in%20the%20context%20of%20analytical%20chemistry&text = Chemometric%20techniques%20are%20ideal%20for%20analyzing%20chemical%20structures%20and%20spectra.&text = A%20widely%20used%20application%20of,and%20spectra%2C%E2%80%9D%20explains%20Lee> Accessed 1.21.

European Commission. (2021). *EU circular economy action plan.* <https://ec.europa.eu/environment/circular-economy/> Accessed 2.21.

Grace, D. (2019). Infectious diseases and agriculture. Reference module in food science. *Encyclopedia of Food Security and Sustainability, 3*, 439−447.

Grześkowiak, T., Czarczyńska-Goślińska, B., & Zgoła-Grześkowiak, A. (2016). Current approaches in sample preparation for trace analysis of selected endocrine-disrupting compounds: focus on polychlorinated biphenyls, alkylphenols, and parabens. *TrAC, Trends in Analytical Chemistry, 75*, 209−226.

Herman, C. (2015). *Agricultural chemistry: New strategies and environmental perspectives to feed a growing global population.* Washington, DC: American Chemical Society.

Hippe, Z. (1983). Problems in the application of artificial intelligence in analytical chemistry. *Analytica Chimica Acta, 150*, 11−21.

International Union of Pure and Applied ChemistryStephenson, G. R., Ferris, I. G., Holland, P. T., & Norberg, M. (2006). Glossary of terms related to pesticides. *Pure and Applied Chemistry, 78*, 2075−2154.

Kumar, V., Guleria, P., & Mehta, S. K. (2017). Nanosensors for food quality and safety assessment. *Environmental Chemistry Letters, 15*, 165−177.

Li, C., Jin, F., & Snyder, S. A. (2018). Recent advancements and future trends in analysis of nonylphenol ethoxylates and their degradation product nonylphenol in food and environment. *TrAC, Trends in Analytical Chemistry, 107*, 78−90.

Liu, J., Ma, K., Ciais, P., & Polasky, S. (2016). Reducing human nitrogen use for food production. *Scientific Reports, 6*, 30104. Available from https://doi.org/10.1038/srep30104.

Mesnage, R., & Antoniou, M. N. (2018). Ignoring adjuvant toxicity falsifies the safety profile of commercial pesticides. *Frontiers in Public Health, 5*, 361. Available from https://doi.org/10.3389/fpubh.2017.00361.

Munawar, A., Ong, Y., Schirhagi, R., Tahir, M. A., Khan, W. S., & Bajwa, S. Z. (2019). Nanosensors for diagnosis with optical, electric and mechanical transducers. *RSC Advances, 9*, 6793−6803.

Organisation for Economic Co-operation and Development. (2018). *Series on pesticides no. 97 and series on testing and assessment no. 279 guidance document on residues in rotational crops.* Paris: OECD.

Organisation of Economic Co-operation and Development. (2021a). *Sustainable chemistry.* <http://www.oecd.org/chemicalsafety/risk-management/sustainablechemistry.htm#:~:text = %22Sustainable%20chemistry%20is%20a%20scientific,for%20chemical%20products%20and%20services> Accessed 1.21.

Organisation of Economic Co-operation and Development. (2021b). *OECD series on principles of good laboratory practice (GLP) and compliance monitoring.* <http://www.oecd.org/chemicalsafety/testing/oecdseriesonprinciplesofgoodlaboratorypracticeglpandcompliancemonitoring.htm> Accessed 2.21.

Pingali, D. (2012). Green revolution: Impacts, limits, and the path ahead. *Proceedings of the National Academy of Sciences of the United States of America, 109*, 12302−12308.

Pinto-Zaervallos, D. M., & Zarbin, P. H. G. (2013). Química na agricultura: Perspectivas para o desenvolvimento de tecnologias sustentáveis [*Chemistry in agriculture: Perspectives for the development of sustainable technologies*]. *Química Nova, 36*, 1509−1513.

Quintero-Angel, M., & González-Acevedo, A. (2018). Tendencies and challenges for the assessment of agricultural sustainability. *Agriculture, Ecosystems & Environment, 254*, 273−281.

Ritter, S. K. (2008). The Harber-Bosh reaction: An early chemical impact on sustainability. *Chemical & Engineering News*. Available from https://pubs.acs.org/cen/coverstory/86/8633cover3box2.html, Accessed 1.21.

Statista. (2020). *Market value of global agrochemicals from 2006 to 2019 by type (in millions of U. S. dollars)*. <https://www.statista.com/statistics/311953/agrochemical-market-value-worldwide-by-type/#:~:text=In%202019%2C%20the%20global%20agrochemical, insecticides%2C%20herbicides%2C%20and%20fungicides> Accessed 1.21.

United Nations Food and Agriculture Organization. (2010). *International code of conduct on the distribution and use of pesticides. Guidelines for the registration of pesticides*. Rome.: FAO/OMS.

U.S. Geological Survey. (2020). *Nitrogen data sheet—mineral commodity summaries 2020*. <https://pubs.usgs.gov/periodicals/mcs2020/mcs2020-nitrogen.pdf> Accessed 2.21.

Vasseghian, Y., Berkani, M., Almomani, F., & Dragoi, E.-N. (2021). Data mining for pesticide decontamination using heterogeneous photocatalytic processes. *Chemosphere, 270*, 129449.

Vaz, Jr. S. (Ed.), (2019). *Sustainable agrochemistry—A compendium of technologies*. Chan: Springer Nature.

CHAPTER 8

Sorption study for environmental purpose

The introduction of antimicrobial agents in therapeutic use for livestock and poultry was revolutionary in combating diseases caused by bacteria in the middle of the last century. In animal feed, it contributed to the considerable increase in animal production. However, this is currently under discussion because of the development of bacterial resistance and its impact on the treatment of diseases in humans. The use of antibiotics in animal production is considered by the World Health Organization an increasing risk for human health (World Health Organization, 2020).It is an example of situation that demands the residue analyses in food and environmental matrices.

The fate of agrochemicals in the environment and their effects are subject of research for academic and regulation purposes. For instance, a certain agrochemical after its application can achieve soil components, be transported, and reach the surface water or groundwater, producing deleterious effects to humans and animals.

These aspects can be evaluated by the use of analytical techniques—especially chromatographic techniques—allied to mathematical models and treatments and considering molecular physicochemical processes.

8.1 Sorption isotherms

The sorption processes, or surface interactions, are a phenomena that promotes the effects of agrochemical molecules on a certain matrix which occur due to physiosorption and chemisorption.

In physiosorption, there are weak van der Waals interactions (e.g., dispersion and dipolar interaction) and hydrogen bonding between the sorbate (agrochemical) and the surface of the sorbent, or substrate (i.e., the analytical matrix). The energy released when a particle is adsorbed is of the same order of magnitude as the condensation enthalpy, around 20 kJ mol^{-1}, with an accommodation of sorbate molecules on the sorbent surface due to the dispersion of thermal movement (Atkins & De Paula, 2006).

In chemisorption, sorbate molecules are sorbed on the surface of the substrate by the formation of chemical bonds, usually covalent bonds, tending to find sites that maximize their number of coordination with the substrate. The enthalpy of chemisorption is greater than that of physiosorption. These values are around 200 kJ mol^{-1} and the distance between sorbate and sorbent is often shorter for chemisorption (Atkins & De Paula, 2006).

Sorption isotherms are curves that represent the concentration of chemical species sorbed by its nonsorbed concentration, usually in an aqueous medium, at a constant temperature. A mathematical model can be assigned to them, with such isotherms being used to quantify the sorption capacity of a sorbent against experimental variables (Perzynski, Sims, & Vance, 2005). According to Giles, Smith, and Huitson (1974), the assignments that can be made on the most common types of sorption isotherms are:

- Isotherm "S": result of cooperation of each molecule successively adsorbed to the adsorption of others, with the availability of intermolecular space limiting sorption;
- Isotherm "C", or partition: result of increased availability of surface interaction, leading to partition.[1]

The linear sorption, or partition, isotherm is defined by Eq. (8.1):

$$K_d = \frac{C_s}{C_e} \qquad (8.1)$$

where C_s is the concentration of the solute absorbed in mmol g^{-1} and C_e is the concentration of the solute in the aqueous phase in equilibrium in mmol L^{-1}, with K_d in L g^{-1}, which is the partition constant, providing the trend of solute in remaining on the sorbent surface or in aqueous phase after reaching chemical balance between species.

The Freundlich isotherm, applied to heterogeneous surfaces, is defined by Eq. (8.2):

$$S = k_f C_e^N \qquad (8.2)$$

---

[1] Partition is a term often used as a synonym for distribution and extraction. However, an essential difference exists by definition between distribution constant or partition ratio and partition constant. The term partition should be, but is not invariably, applied to the distribution of a single definite chemical species between the two phases. Source: IUPAC. *Pure and Applied Chemistry (1993)*, v. 65. (*Nomenclature for liquid-liquid distribution (solvent extraction) (IUPAC Recommendations 1993)*) on page 2378.

where S is the concentration of the sorbed solute in mg g^{-1}, k_f is the sorption coefficient in mg^{1-n} L × g^n, Ce is the concentration of the solute in equilibrium in aqueous solution in mg L^{-1} and N is the linearity parameter. In a simple case, $N = 1$ and k_f is equivalent to K_d for linear isotherms. The logarithmic form of the equation is used for the linear adjustment of the model [Eq. (8.3)]:

$$\text{Log}S = \text{log}K_f + N\text{log}C_e \tag{8.3}$$

The Langmuir isotherm, which is restricted to homogeneous surfaces, is represented by Eq. (8.4):

$$S = \frac{C_m C_e}{K_L + C_e} \tag{8.4}$$

where S is the concentration of the solute adsorbed in mg g^{-1}, C_m is the maximum amount of the solute adsorbed in mg g^{-1}, C_e is the concentration of the solute in equilibrium in aqueous solution in mg L^{-1} and K_L is the coefficient of Langmuir sorption in mg L^{-1}.

As mentioned above (Giles et al., 1974), they also classified sorption isotherms based on their shapes and curvatures. When taking into account plateaus, inflection and maximum points, subgroups of isotherms were also distinguished. However, this operational classification is purely empirical and does not reveal the processes that led to the different forms of isotherms (Hinz, 2001). Table 8.1 presents these classes of isotherms and their characteristics according to the mathematical treatment applied. In reference to the characteristics of the sorbent-sorbate interaction, and according to comments already made regarding sorption isotherms, these curves can be interpreted as follows (Perzynski et al., 2005):

- "L" curve: represents a sorbate that has a high affinity for the sorbent at a low concentration, but at a high concentration this affinity decreases, together with the number of sorption sites;
- "H" curve: represents a sorbate with a high affinity for the sorbent;
- "S" curve: suggests that there is a barrier to sorbate sorption, but once this limitation is overlapped, sorption is similar to that of L, that is, sorption increases with the increase in the concentration of sorbate in the aqueous phase;
- "C" or linear curve: partition characteristic involving organic compounds and soil organic matter (SOM), suggests that the sorbate retention depends on the sorbent's surface coating, that is, the sorbate has greater affinity for the surface sorbent than the water phase.

Table 8.1 Forms of sorption isotherms proposed by Giles et al. (1974).

Class	Subgroup	Isotherm Forms ($S \times C_e$)	Isotherm Forms ($K_d \times S$)	Isotherm Forms ($\log K_d \times \log S$)
S	1	Concave	Positive slope	Negative slope
L	1	Convex	Negative slope,	Negative slope
H	1	Convex	Negative slope,	Negative slope
C	1	Straight line	Zero slope	Zero slope
S, L, H	2	Plateau	Negative slope, line or curve	$S_T = S$ for reduced values for K_d
S, L, H	3	Plateau, inflection	Minimum	Minimum
S, L, H	4	Plateau, inflection,	Minimum	Minimum

Hinz, C. (2001) Description of sorption data with isotherm equations. *Geoderma, 99*, 225–243. S is the adsorbed concentration, C_e is the equilibrium concentration, and K_d is the partition constant or if sorption.

Sorption isotherms can be used to:
- Verify the sorption between soils and/or SOM at different pH values with the use of HPLC to construct isotherms to quantify the sorption capacity;
- Obtain information that can assist in the understanding of agrochemical transport in soil;

- Contribute to the input of data to assess the environmental chemical risk related to the large-scale use of chemicals for agricultural and for veterinary purposes.

Studies of sorption are used for registration (i.e., physicochemical assays) (Organisation for Economic Cooperation and Development, 2000) and for environmental fate (Liang, Lui, & Allen, 2018) purposes.

8.2 Methodology

Fig. 8.1 illustrates the strategy to be applied for sorption studies in soils.

8.2.1 Soil characterization

To apply the strategy in Fig. 8.1 two soils from Brazilian tropical region in the São Paulo State were utilized as described in the following.

Soils M1 (forest, red-yellow dystrophic oxisol—21040′4″S and 47050′33″O) and T1 (peat, organosol—21033′20″S and 47055′08″O) were collected in superficial depth (0−15 cm) and later used in a sorption experiment. Table 8.2 describes the physicochemical and physical parameters used in the characterization, in which they will supply additional information to understand the molecular interaction.

The soil samples were dried at room temperature for 7 days. After this period, the samples were homogenized in a 2 mm granulometric sieve and sent to the laboratory for analyses.

8.2.2 Sorption soils-agrochemical

For the study, the veterinary antibiotic oxytetracycline (OTC, Fig. 8.2) was chosen as the agrochemical molecule and source of chemical residue, due its large use in livestock and poultry and the suspect to produce antibiotic resistance in the environment (Vaz Jr, Lopes, & Martin-Neto, 2015).

A factor of concerning in the use of veterinary tetracycline antibiotics is that the animal organism, does not absorb a large fraction of them being excreted between 20% and 90% unmodified by means of urine and feces (Chee-Sandford, Aminov, Krapac, Garrigues-Jeanjean, & Mackie, 2001). This can lead to the developing of resistance to these antibiotics, and depending on their transport in the soil, reach surface water or groundwater and provide resistance by water microorganisms. If humans and animals consume the water, microorganisms will be resistant in the recipient

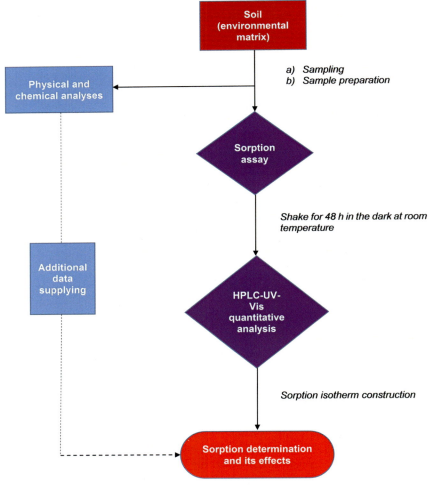

Figure 8.1 Flowchart of the sorption determination method for a certain agrochemical molecule to soils. *Credit: author.*

organism to medication by these antibiotics, being able to transmit genes of resistance to such drugs (Boxall et al., 2006; Liou et al., 2009). The application of swine, bovine, and poultry manure to cultivable areas stands out as a source of soil pollution by antibiotics (Venglovsky, Sasakova, & Placha, 2009), with evidence of the presence of their residues in the soil resulting from this type of agricultural application (Martínez-Caballo, Gonzáles-Barreiro, Scharf, & Gans, 2007; Wang & Yates, 2008). Added to this environmental and public health problem is the fact that the new

Table 8.2 Analytical parameters of soil characterization and their use for the interpretation of sorption data.

Analytical parameter	Purposes of use	Technique
pH at CaCl$_2$ solution	To investigate the influence of protonation/deprotonation for the interaction	Direct potentiometry
Al^{3-}	Determine the availability of trivalent metal cations for bonding	Atomic absorption spectrometry
H$^+$ + Al^{3+}	Determine the availability of trivalent metal cations and protons for bonding	Titulometry
Ca^{2+}	Determine the availability of bivalent metal cations for bonding	Atomic absorption spectrometry
Mg^{2+}	Determine the availability of bivalent metal cations for bonding	Atomic absorption spectrometry
K$^+$	Determine the availability of monovalent metal cations for bonding	Atomic absorption spectrometry
Base saturation	Determine the percentage of soil cation exchange sites occupied by bases	Titulometry
Sum of bases	Determine the total availability of negative sites for binding, which are occupied by alkaline and alkaline earth metals (e.g., Na$^+$, K$^+$, Ca^{2+}, Mg^{2+})	Titulometry
Organic matter (OM)	Determine the presence and the origin of the OM for interaction and its influence on this interaction	Gravimetry
Cation exchange capacity (CEC, total)	Determine the maximum amount of negative charges to be released at pH 7 value	Cationic exchange column
Granulometry (sand, silt and clay)	Classify the inorganic composition of the soil in terms of particle size	Sifting

Parameter definition and analytical methods adapted from, Brazilian Agricultural Research Corporation. (1997). *Geoquímica de alguns solos brasileiros [Geochemistry of some Brazilian soils]*. Rio de Janeiro: Embrapa; Brazilian Agricultural Research Corporation. (1998). *Análises químicas para avaliação da fertilidade do solo: métodos usados na Embrapa Solos [Chemical analysis for soil fertility assessment: Methods used at Embrapa Soils]*. Rio de Janeiro: Embrapa; Brazilian Agricultural Research Corporation. (1999). *Manual de análises químicas de solos, plantas e fertilizantes [Manual of chemical analysis of soils, plants and fertilizers]*. Brasília: Embrapa.

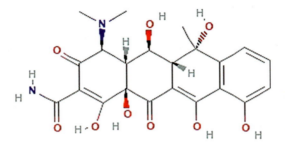

Figure 8.2 The oxytetracycline hydrochloride chemical 2D-structure. *Reprinted with permission from Pubchem. Oxytetracycline hydrochloride. (2021). <https://pubchem.ncbi.nlm.nih.gov/compound/54680782> Accessed 2.21.*

generation of growth promoters—antibiotics used to quench nutrient competition by microorganisms in the digestive system—is not absorbed by the intestinal tract, to avoid residues in the meat (Andrade, 2007), which may increase the presence of these compounds in environmental matrices. The use of these promoters can lead to biochemical and microbiological risks, since studies carried out in Brazil have shown that broilers can act as a reservoir of genes resistant to antibiotics of important use in veterinary and human medicine, including tetracyclines (Pessanha & Gontijo-Filho, 2001)—it was observed also in Europe (Blanco, Lemus, & Grande, 2009).

The physical, chemical, and structural properties of OTC, in its form of oxytetracycline chlorohydrate, are shown in Table 8.3.

The soils-OTC sorption study was carried out in duplicate using the soils M1 and T1 according to the adaptation of the procedure developed by Jones, Bruland, Agrawal, and Vasudevan (2005). As the load and speciation of OTC and soil are dependent on pH value, a buffer solution was selected for pH value of 4.8, close to the maximum mass fraction of the OTC^{+-} zwitterion and in order to minimize possible interferences of pH values and electronic charges for sorption.

OTC solutions were previously prepared in the following concentrations in a buffer solution sodium acetate/acetic acid 0.1 mol L^{-1} in ultrapure water and from an OTC stock solution at 240 mg L^{-1} from oxytetracycline hydrochloride with 95% of purity (minimum) and HPLC grade: 120, 60, 30, 20, 10 and 5 mg L^{-1}; a 0 mg L^{-1} control solution was

Table 8.3 Physicochemical properties of oxytetracycline chlorohydrate.

CAS number	2058-46-0
Empirical molecular formula	$C_{22}H_{24}N_2O_9 \cdot HCl$
Molecular mass	496,89 g mol^{-1}
H donors for hydrogen bonding	8
H acceptors for hydrogen bonding	11
Charge	0
Covalent bonding units	2
Polar surface area	202
Purity	$\geq$ 95% (HPLC grade)
Solid state shape	Crystals
Color	Yellow
Water solubility	0.6–0.8 mg mL^{-1}

Adapted from Sigma-Aldrich. *Oxytetracycline hydrochloride.* (2021). <https://www.sigmaaldrich.com/catalog/product/sigma/o5875?lang = pt&region = BR> Accessed 2.21 and Pubchem. *Oxytetracycline hydrochloride.* (2021). <https://pubchem.ncbi.nlm.nih.gov/compound/54680782> Accessed 2.21.

maintained. A 10 mL aliquot of each solution was transferred to an amber glass flask and immediately afterwards soil mass was added to a concentration of 5 g L^{-1}. The mixtures were stirred on a rotary table for 48 hours, protected from light and at room temperature, and after this period they proceeded to chromatographic analysis and the results obtained were used in the construction of the sorption isotherms.

For HPLC analysis, a high-performance liquid chromatograph with UV-Vis detector, and a column of reverse stationary phase of the type PSDVB (polystyrene divinylbenzene), 15 cm long, 4.6 mm internal diameter, and 5 μm of particle size was used, with the following operating conditions adapted from Loke, Tjørnelund, and Halling-Sørensen (2002):

- Isocratic mobile phase (v/v): 26% acetonitrile/74% aqueous trifluoroacetic acid solution at 0.05% (v/v), with the liquid solutions being previously deaerated in an ultrasound bath for 15 minutes to avoid the presence of air bubbles in the system;
- Column temperature: 30°C;
- Mobile phase flow: 1 mL min^{-1};
- Wavelength of the UV-Visible detector: 355 nm.

The acetonitrile used was of HPLC grade while the trifluoroacetic acid was 99% pure.

An external OTC calibration curve was constructed for the analytical method and the points were 5, 10, 20, 30, 60, 120, and 240 mg L^{-1}. From this curve, the LOQ of the method was determined.

8.3 Results and discussion
8.3.1 Soil characterization
Table 8.4 shows the results of the soil characterizations.

The data presented demonstrated that the soil T1 had a greater amount of OM and metals, which was reflected in higher values of pH and CEC, respectively, when compared to soil M1.

8.3.2 Sorption soil-agrochemical
Isotherms were built to check the influence of soil type on OTC sorption, choosing soils with different characteristics.

Firstly, it should be demonstrated that the external calibration curves of the analytical method, shown in Fig. 8.3, obtained values of the linear correlation coefficient R of 0.999, which were satisfactory for analytical purposes (Ribani, Bottoli, Collins, Jardim, & Melo, 2004).

LOQ was determined according to Eq. (8.5) (Ribani et al., 2004), following IUPAC recommendations (International Union of Pure and Applied Chemistry, 2021):

$$\text{LOQ} = 10 \times \frac{s}{S} \tag{8.5}$$

Table 8.4 Characterization results for M1 (forest, red-yellow dystrophic oxisol) and T1 (peat, organosol) soils.

Analytical parameter	Unity	Soil M1	Soil T1
pH at CaCl$_2$ solution	—	4	6.3
Al^{3+}	Mmol L^{-1}	8	0
H$^+$ + Al^{3+}	Mmol L^{-1}	72	15
K$^+$	Mmol L^{-1}	0.5	6.3
Ca^{2+}	Mmol L^{-1}	4	80
Mg^{2+}	Mmol L^{-1}	2	42
CEC (total)	Mmol L^{-1}	79	143
OM	g L^{-1}	26	119
Sum of bases	Mmol L^{-1}	7	128
Base saturation	—	8%	90%
Granulometry	—		
Sand		4%	19.2%
Silt		78%	19.2%
Clay		18%	61.6%

Sorption study for environmental purpose 227

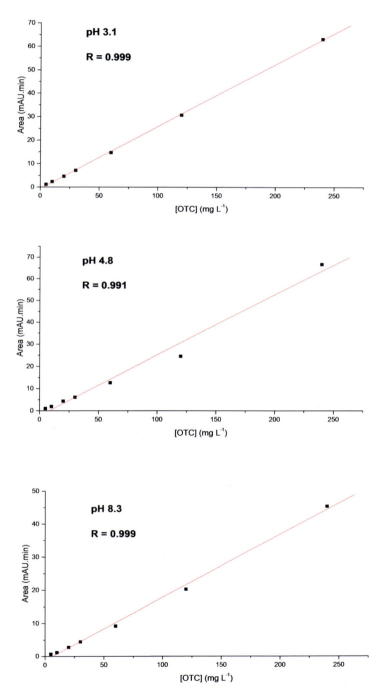

Figure 8.3 Calibration curves (external standard) at different pH values for the chromatographic method of oxytetracycline (OTC) determination. *Credit: author.*

Where s is the estimate of the standard deviation of the response, in this case the linear regression equation, and S is the slope or slope of the analytical curve. Thus, a LOQ of 4.0 mg L^{-1} was determined for the three curves in Fig. 8.3, with the method not showing any major variations depending on the pH value of the medium.

Chromatograms in Figs. 8.4 and 8.5 present the chromatographic profiles of the soil-OTC sorption products.

For the initial OTC concentrations of 10.0 and 5.0 mg L^{-1}, negative values for C_e were obtained, which demonstrated that the LOQ was inefficient for these two concentrations, most likely due to the decreased sensitivity of the method.

Linear or partition isotherms, and Freundlich and Langmuir isotherms were tested, with subsequent evaluation of the model that best suited after linear regression, or linear adjustment, when relevant for both mathematical models (Freundlich and Langmuir). These isotherms are shown in Figs. 8.6 and 8.7.

According to Fig. 8.6, the linear or partition isotherm was not followed, and it is necessary to test the Freundlich and Langmuir isotherms (Figs. 8.7 and 8.8).

It should be noted that the N value of the Freundlich isotherm allows to infer about the shape of the isotherm and the sorption mechanism, where $N=1$ represents a linear or partition C isotherm, $N<1$ a L isotherm and $N>1$ a S isotherm (Hinz, 2001). As observed by Ferreira,

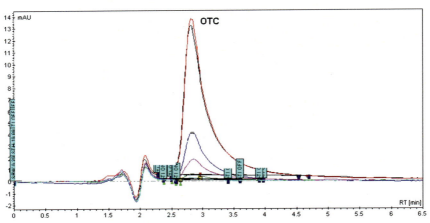

Figure 8.4 Overlapped chromatograms for soil "M1"-OTC obtained at pH 4.8, where: C_i OTC = 120 mg L^{-1} (black), C_i OTC = 60 mg L^{-1} (red), C_i OTC = 30 mg L^{-1} (blue), C_i OTC = 20 mg L^{-1} (pink). Credit: author.

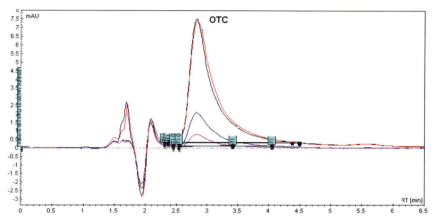

Figure 8.5 Overlapped chromatograms for soil "T1"-OTC obtained at pH 4.8, where: C$_i$ OTC = 120 mg L^{-1} (*black*), C$_i$ OTC = 60 mg L^{-1} (*red*), C$_i$ OTC = 30 mg L^{-1} (*blue*), C$_i$ OTC = 20 mg L^{-1} (*pink*). *Credit: author.*

Martin-Neto, Vaz, and Reginato (2002), the higher the N value, the more heterogeneous the sorption process is.

From the linear adjustment of the obtained curves, it was determined that the Freundlich isotherm, in the linear logarithmic form adjusted with the use of the ORIGIN software,[2] was the model that best suited. It should be noted that this isotherm is commonly used for the empirical representation of sorption of organic compounds to heterogeneous surfaces (e.g., soil and SOM), as veterinary drugs (Ferreira et al., 2002; Yu, Fink, Wintgens, Melin, & Ternes, 2009). Table 8.5 shows the fitting parameters.

The values obtained for K_f demonstrated that the soil with the highest OM content (T1) presented the highest sorption capacity, with an L-type sorption, based on N, which increases with the increase in the concentration of OTC in the aqueous medium (Perzynski et al., 2005). The values obtained for K_f were below those obtained in the literature for OTC in temperate soils (Jones et al., 2005), demonstrating that this sorption was weak. However, Jones et al. (2005) observed that the partition constant K_d is more suitable for the observation of the sorption capacity than K_f. In this way, K_d was calculated for the two soils (Table 8.6).

Fig. 8.9 presents the graph with the K_d profile for each soil.

The behavior in Fig. 8.9 can be interpreted from the composition characteristics of each soil. Jones et al. (2005) observed that soil texture

---

[2] https://www.originlab.com/

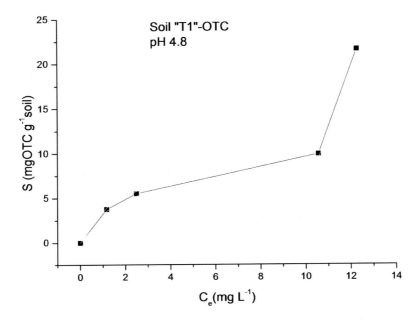

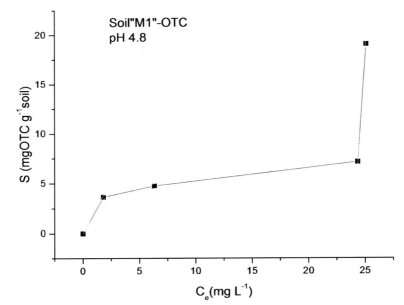

Figure 8.6 Isotherms obtained for soil-oxytetracycline (OTC), according to the linear or partition model. *Credit: author.*

Sorption study for environmental purpose 231

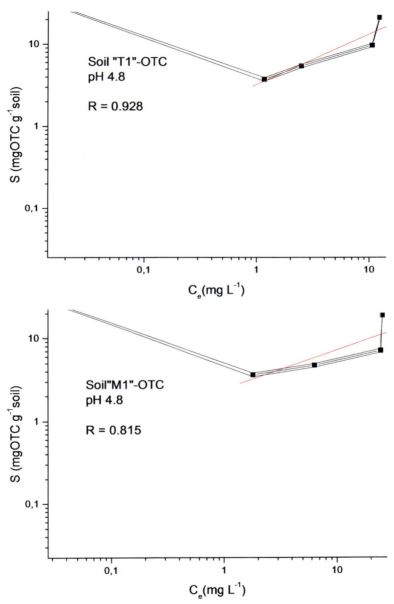

Figure 8.7 Isotherms obtained for soil-oxytetracycline (OTC), according to the Freundlich model after linear adjustment. *Credit: author.*

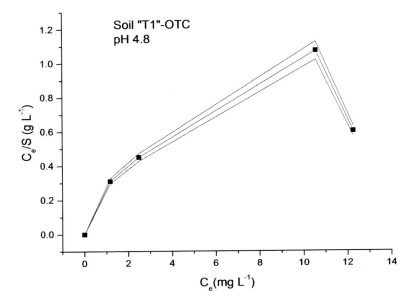

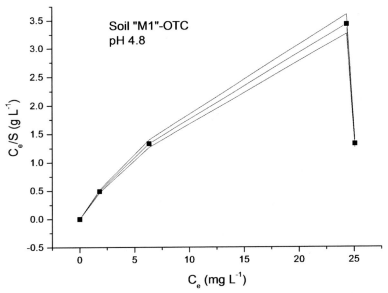

Figure 8.8 Isotherms obtained for soil-oxytetracycline (OTC), according to the Langmuir model. *Credit: author.*

Table 8.5 Values calculated for Freundlich isotherms after linear fitting of the equation in its logarithmic form.

Freundlich Isotherm Equation	K_f (mg^{1-n} L g^n)	N	Isotherm form	R	Soil
Log S = 0.51 + 0.62logC$_e$	3.24	0.62	L	0.928	T1
Log S = 0.39 + 0.47logC$_e$	2.45	0.47	L	0.815	M1

Table 8.6 Values determined for K_d for sorption of OTC to soils T1 and M1.

C$_i$ OTC (mgL^{-1})	K_d (Lkg^{-1})	Soil
120.0	9290	T1
	760	M1
60.0	940	T1
	290	M1
30.0	2210	T1
	750	M1
20.0	3220	T1
	2020	M1

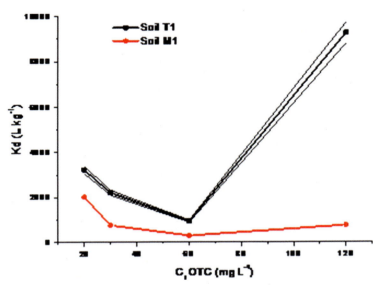

Figure 8.9 Overlapping of K_d profiles as a function of the initial concentration of oxytetracycline (OTC) for the two soils studied. *Credit: author.*

(granulometry), together with CEC, are the main influencing factors for OTC sorption in soils. However, such influence is limited to the content of OM present, which decreases with the increase of the latter.

When correlating the sorption profiles presented in Fig. 8.9 with the characterization of the soils in Table 8.6, it was observed that the soil with the highest OM content and the highest CEC (T1) was the one that presented the highest K_d value for the highest initial concentration of OTC (120 mg L^{-1}). A fact to be considered also for the highest K_d value of T1 is the greater presence of metallic cations leads to a greater interaction between OTC and OM by means of metallic bonding, as OTC is a strong chelating agent for metal cations in medium with a pH value around 5 (Couto, Montenegro, & Reis, 2000; Loke et al., 2002).

However, according to values established by the Brazilian environmental agency—IBAMA[3]—for the evaluation of the fate for chemical substances in soils (Brasil, 1990), K_d values between 0 and 24 L kg^{-1} characterize a low sorption capacity and greater transport in the soil and groundwater reach by leaching process, the opposite being observed for soil-OTC sorption. The K_d values in Table 8.6 were close to those obtained in the literature for soils in other regions of the world—between 486 and 12,048 L kg^{-1} (Jones et al., 2005). However, such determined values expressed typical OTC behavior in relation to Brazilian soils.

It is important to highlight that when obeying the Freundlich isotherm, the sorption to soils was due to the formation of multilayers (Atkins & De Paula, 2006), of the sorbed species (OTC) on the surface of the sorbent species (soil).

8.4 Conclusion

Sorption isotherm is not a simple tool for environmental risk assessment. However, it can supply valuable data regarding the agrochemical fate across a certain matrix. As demonstrated, frequently is necessary the fitting of mathematical plots and models in order to achieve the best information.

From the characterization of two soils in the State of São Paulo with different levels of OM, the sorption of OTC was verified in Brazilian soils. It is dependent on the content of OM present, presenting a potential risk to the environment, resulting from its strong sorption capacity to such soils, which makes it difficult to transport by leaching and, consequently,

---

[3] https://www.gov.br/ibama/pt-br

less supply of antibiotics in groundwater. These observations also provide subsidies for further assessment of the chemical risk of the antibiotic to microorganisms and soil activity.

OTC sorption to both soils was in accordance with the Freundlich isotherm (L-type). Such isotherms showed a sorption of homogeneous characteristic for these soils.

Finally, the initially proposed experimental strategy can be replicated to another agrochemical molecules in soils.

References

Andrade, N. A. (2007). Mitos e verdades sobre o uso de antibióticos nas rações [*Myths and truths about the use of antibiotics in feed*]. *Jornal do Conselho Regional de Medicina Veterinária—RJ, 186*, 4−5.

Atkins, P., & De Paula, J. (2006). *Atkins's physical chemistry* (8th ed., pp. 916−922). Oxford: Oxford University Press.

Blanco, G., Lemus, J. A., & Grande, J. (2009). Microbial pollution in wildlife: Linking agricultural manuring and bacterial antibiotic resistance in red-billed choughs. *Environmental Research, 109*, 405−412.

Boxall, A. B. A., Johnson, P., Smith, E. J., Sinclair, C. J., Stutt, E., & Levy, L. S. (2006). Uptake of veterinary medicines from soils into plants. *Journal of Agricultural and Food Chemistry, 54*, 2288−2297.

Brasil (1990). *Ministério do Meio Ambiente. Ibama. Manual de testes para avaliação da ecotoxicidade de agentes químicos [Guidelines for assessing the ecotoxicity of chemical agents]*. Brasília: Ibama.

Brazilian Agricultural Research Corporation (1997). *Geoquímica de alguns solos brasileiros [Geochemistry of some Brazilian soils]*. Rio de Janeiro: Embrapa.

Brazilian Agricultural Research Corporation (1998). *Análises químicas para avaliação da fertilidade do solo: métodos usados na Embrapa Solos [Chemical analysis for soil fertility assessment: Methods used at Embrapa Soils]*. Rio de Janeiro: Embrapa.

Brazilian Agricultural Research Corporation (1999). *Manual de análises químicas de solos, plantas e fertilizantes [Manual of chemical analysis of soils, plants and fertilizers]*. Brasília: Embrapa.

Chee-Sandford, J. C., Aminov, R. I., Krapac, I. J., Garrigues-Jeanjean, N., & Mackie, R. I. (2001). Occurrence and diversity of tetracycline resistance genes in lagoons and groundwater underlying two swine production facilities. *Applied and Environmental Microbiology, 67*, 1494−1502.

Couto, C. M. C. M., Montenegro, C. B. S. M., & Reis, S. (2000). Complexação da tetraciclina, da oxitetraciclina e da clortetraciclina com o catião cobre (II). Estudo potenciométrico [*Complexation of tetracycline, oxytetracycline and chlortetracycline with the copper (II) cation. Potentiometric study*]. *Química Nova, 23*, 457−460.

Ferreira, J. A., Martin-Neto, L., Vaz, C. M. P., & Reginato, J. B. (2002). Sorption interactions between imazaquin and a humic acid extracted from a typical Brazilian oxisol. *Journal of Environmental Quality, 31*, 1665−1670.

Giles, C. H., Smith, D., & Huitson, A. (1974). A general treatment and classification of the solute adsorption isotherm. I. Theoretical. *Journal of Colloid and Interface Science, 47*, 755−765.

Hinz, C. (2001). Description of sorption data with isotherm equations. *Geoderma, 99*, 225−243.

International Union of Pure and Applied Chemistry (2021) *Gold book*. <https://goldbook.iupac.org/> Accessed 2.21.

Jones, A. D., Bruland, G. L., Agrawal, S. G., & Vasudevan, D. (2005). Factors influencing the sorption of oxytetracycline to soils. *Environmental Toxicology and Chemistry, 24*, 761−770.

Liang, Y., Lui, X., & Allen, M. R. (2018). Measuring and modeling surface sorption dynamics of organophosphate flame retardants on impervious surfaces. *Chemosphere, 193*, 754−762.

Liou, F., Ying, G.-G., Tao, R., Zhao, J.-L., Yang, J.-F., & Zhao, L.-F. (2009). Effects of six selected antibiotics on plant growth and soil microbial and enzymatic activities. *Environmental Pollution, 157*, 1636−1642.

Loke, M.-L., Tjørnelund, J., & Halling-Sørensen, B. (2002). Determination of the distribution coefficient (logK_d) of oxytetracycline, tylosin A, olaquindox and metronidazole in manure. *Chemosphere, 48*, 351−361.

Martínez-Caballo, E., Gonzáles-Barreiro, C., Scharf, S., & Gans, O. (2007). Environmental monitoring study of selected veterinary antibiotics in animal manure and soils in Austria. *Environmental Pollution, 148*, 570−579.

Organisation for Economic Cooperation and Development (2000). *OECD guidelines for the testing of chemicals, section 1—Physical-chemical properties. Test. No 106: Adsorption-desorption using a batch equilibrium method*. Paris: OECD.

Perzynski, G. M., Sims, J. T., & Vance, G. F. (2005). *Soil and environmental quality* (3rd ed.). Boca Ranton: CRC Taylor & Francis.

Pessanha, R. P., & Gontijo-Filho, P. P. (2001). Uso de antimicrobianos como promotores de crescimento e resistência em isolados de *Escherichia coli* e de Enterobacteriaceae lactose-negativa da microflora fecal de frangos de corte [*Use of antimicrobials as growth and resistance promoters in isolates of Escherichia coli and lactose-negative Enterobacteriaceae from the fecal microflora of broilers*]. *Arquivo Brasileiro de Medicina Veterinária e Zootecnica, 53*, 111−115.

Pubchem *Oxytetracycline hydrochloride*. (2021). <https://pubchem.ncbi.nlm.nih.gov/compound/54680782> Accessed 2.21.

Ribani, M., Bottoli, C. B., Collins, C. H., Jardim, I. C. S. F., & Melo, L. F. (2004). Validação em métodos cromatográficos e eletroforéticos [*Validation in chromatographic and electrophoretic methods*]. *Química Nova, 27*, 771−780.

Sigma-Aldrich *Oxytetracycline hydrochloride*. (2021). <https://www.sigmaaldrich.com/catalog/product/sigma/o5875?lang = pt&region = BR> Accessed 2.21.

Vaz Jr, S., Lopes, W. T., & Martin-Neto, L. (2015). Study of molecular interactions between humic acid from Brazilian soil and the antibiotic oxytetracycline. *Environmental Technology & Innovation, 4*, 260−267.

Venglovsky, J., Sasakova, N., & Placha, I. (2009). Pathogens and antibiotic residues in animal manures and hygienic and ecological risks related to subsequent land application. *Bioresource Technology, 100*, 5386−5391.

Wang, Q., & Yates, S. R. (2008). Laboratory study of oxytetracycline degradation kinetics in animal manure and soil. *Journal of Agricultural and Food Chemistry, 56*, 1683−1688.

World Health Organization (2020) *Antibiotic resistance*. <https://www.who.int/news-room/fact-sheets/detail/antibiotic-resistance#:~:text = Bacteria%2C%20not%20humans%20or%20animals,hospital%20stays%2C%20and%20increased%20mortality> Accessed 2.21.

Yu, L., Fink, G., Wintgens, T., Melin, T., & Ternes, T. A. (2009). Sorption behavior of potential organic wastewater indicators with soils. *Water Research, 43*, 951−960.

CHAPTER 9

A practical guide for residue analysis

An analytical laboratory is a very sophisticated workplace with highly qualified people. Despite the cutting-edge technical and scientific knowledge located there, its management in order to generate reliable results and profits should follow concise guidelines of those tasks to be performed taking into account challenges to be overcome daily. Then it is import to revisit some fundamentals seen in the previous chapters; however, with more targeting and applicability.

9.1 Establishing (indeed) a quality control

A quality control (QC) police should be applied in order to improve quality and traceability for materials (samples and reference materials) and analytical processes (the analysis itself), which will impact the final analytical result.

It is not always easy to establish a QC policy and plan in an analytical laboratory for residues, despite the globally preconized norms for QC and quality assurance (QA), already dealt with in Chapter 4: Fundamentals of Analytical Chemistry. Some difficulties are related to:
1. Budget availability.
2. Insufficient or inadequate infrastructure.
3. Absence of a team—or a management—dedicated to quality management.
4. Quality procedures overlapping the routine analysis processes.

Then it is very important get a pragmatic view of the laboratory board in order to establish a feasible QC. That means some questions should be answered:
1. What we need? ⇒ Good laboratory practices, ISO 17025, etc.
2. How to make it? ⇒ Hiring a consultant.
3. How to achieve it? ⇒ Hard team training.

In a residue analysis laboratory, QA is achieved by having total and absolute control over all stages of the analytical process, including the preanalytical (i.e., sampling), analytical (i.e., analysis itself) and postanalytical phases (i.e., data interpretation and report release).

Quality management, in turn, encompasses the actions used to produce, direct, and control—that is, QC—this quality, including the determination of a quality policy and objectives, using indicators and targets. The QA of all phases can be achieved through the standardization of each activities involved, from customer service to the release of the report. With that, we can achieve the quality that we want and, with quality management, guarantee it.

It is valuable to take into account that with the improved quality, losses can be avoided, reducing costs, and increasing productivity, and with that, there will be an improvement in market competitiveness.

As a suggestion of permanent international forums for quality in analytical chemistry to be consulted:
1. EURACHEM: a European forum that promotes best practices in analytical measurement by producing authoritative guidance within its expert working groups, publishing guides on the web, and supporting workshops to communicate good practice (Eurachem, 2021).
2. Cooperation on International Traceability in Analytical Chemistry (CITAC): an intercontinental organization that aims to foster collaboration between existing organizations to improve the international comparability of chemical measurements (Cooperation on International Traceability in Analytical Chemistry, 2021).

9.2 Avoiding problems in the sampling step

The standardization of sampling methods is a gap to be fulfilled because it is a source of certain problems during this pivotal initial step, as incorrect collection of representative portion, nonrepresentative portion, and incorrect sample identification generating an incorrect result, among others.

Sampling an agricultural raw material or product should result in a sample that preserves the original characteristics, so some care must be observed to guarantee this requirement. It comprises:
1. Regarding origin and conservation:
   a. Check the identity of the product to be sampled regarding: origin, brand, classification, lot, validity, among others.
   b. For packaged products, check the inviolability of the packaging.
   c. Do not collect products that are in inadequate storage conditions and conservation.

2. Regarding collection and handling:
    a. Do not collect damaged product.
    b. Do not wash the product.
3. Use a disposable glove or plastic bag (suitable for food), so that the hands do not come into direct contact with the product to be collected, avoiding contamination with possible agrochemical residues (e.g., pesticides) from contact, which may compromise the next product to be sampled. To the collection of the next product, use a new glove or plastic.
4. Handle samples carefully to avoid possible damage and removal of surface residues.
    a. Do not transport the samples with other chemicals.

9.2.1 Chain of custody

The chain of custody is the document where sample identity, origin, conditions of storage, responsible analyst, among other information, are stated for traceability control and compliance. Fig. 9.1 depicts an example of a chain of custody.

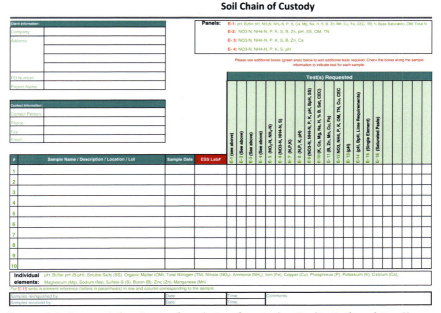

Figure 9.1 An excel factsheet of a chain of custody. *Credit: Author. https://www.microsoft.com/pt-br/microsoft-365/excel.*

9.3 Laboratory management

As preconized by the quality policies, all of the laboratory activities must be documented using standard operating procedures (SOP) or work instructions (IT), which must always be accessible to employees involved in the activities.

A laboratory information management system (LIMS) is a software tool dedicated to manage laboratory data, instruments, and workflows in order to improve lab productivity and efficiency. It keeps track of data associated with samples, experiments, laboratory workflows, and instruments. A LIMS acts as an additional member of the team, automating workflows and tracking all the important sample information, data, workflows, and QA/QC results generated each day. A modern LIMS has evolved from a sample-based tracking system to the digital backbone of the laboratory. It's a tool that helps manage efficiency and costs. As cited by Thermo Fisher Scientific (2021), LIMS does more than just keep track of the sample information, scientific data, and results. It allows to actively manage the entire laboratory process from instrument maintenance and samples to people and consumables. A LIMS manages laboratory samples and associated data, standardizes workflows, reduces human error, and increases efficiency. Furthermore, LIMS adhere to SOPs and maintain high quality and reproducible results.

Nowadays, there are several LIMS suppliers with their products available to purchase according to the customers' necessities.

9.4 Reporting and interpreting analytical results

A critical view and review are paramount to achieve the best result reported from the best result interpretation. That means the analytical chemist always doubts the result in order to guarantee its reliability. From this premise the critical evaluation of a result—for example, the assignments of a certain retention time—should be conducted taking into account the figures of merit deeply discussed in Chapter 4: Fundamentals of Analytical Chemistry, such as accuracy, linearity, limit of detection (LOD), limit of quantification (LOQ), precision, selectivity, sensitivity, and robustness. It is worth to mention that the relevance of good statistics knowledge for the best analytical approach.

For instance, Fig. 9.2 illustrates the concept of univariate decision limit and univariate LOD of single-component calibration for univariate data

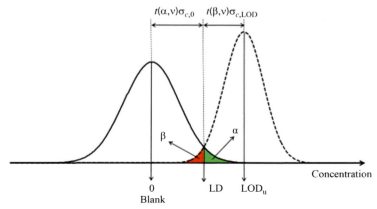

Figure 9.2 Illustration of the IUPAC definition of the univariate decision limit (LD) and limit of detection (LOD$_u$). Two Gaussian bands are centered at the blank and at the LOD$_u$, respectively. The LD helps to decide whether the analyte is detected or not with a rate α of false detects, whereas the LOD$_u$ implies detection with a rate α of false detects and a rate β of false nondetects. The shaded areas correspond to the rate of false detects (*green*) and false nondetects (*red*). *Olivieri, A. C. (2015). Practical guidelines for reporting results in single- and multi-component analytical calibration: A tutorial. Analytica Chimica Acta, 868, 10–22. Reprinted with permission from Elsevier.*

(Olivieri, 2015). It can help to detect α-errors or type I errors (rejection of a true null hypothesis or false positives) (Currie, 1995).

The adequate analytical report should present not only the result itself, but supporting data based on figures of merit as:
1. Calibration curves.
2. Method performance.
3. Method linearity.
4. LOD.
5. LOQ.
6. Uncertainty.

Among others when necessary.

9.5 Estimation of measurement uncertainty of results

According to documents dedicated to chemical metrology as the EURACHEM/CITAC Guide to Quality in Analytical Chemistry (Barwick, 2016), measurement uncertainty characterizes the range of values, within which the real value must lie, with a specified level of confidence. Each measure has an uncertainty associated with it, resulting from errors originating from the various stages of sampling and analysis, and from imperfect knowledge of factors affecting the result.

For the measures to be of practical value, it is necessary to have some knowledge of their reliability or uncertainty. A statement of the uncertainty associated with a result conveys to the customer the quality of the result.

An uncertainty statement is a quantitative estimate of the limits, within which the value of a measurand (such as an analyte concentration) is expected to fall. The uncertainty can be expressed, as a standard deviation or a calculated multiple of the standard deviation (seen in the Chapter 4, Fundamentals of Analytical Chemistry). In obtaining or estimating the uncertainty related to a specific method and analyte, it is essential to ensure that the estimate explicitly considers all possible sources of uncertainty, and to evaluate significant components. Repeatability or reproducibility, for example, is not usually a complete estimate of uncertainty, as none of them fully takes into account any uncertainties associated with systematic effects inherent in a method.

The size of the uncertainty contributions can be estimated in several ways. The value of an uncertainty component associated with random variations in influencing factors can be estimated by measuring the dispersion in results over an appropriate number of determinations under a range of representative conditions—in such an investigation, the number of measurements should normally not be less than ten. The components of uncertainty originating from imperfect knowledge, for example, from a potential trend or trend, can be estimated based on a mathematical model, informed professional judgment, comparisons between international laboratories, experiments on model systems, etc. These different methods for estimating the individual components of uncertainty may be valid.

Often it is not necessary to assess the uncertainties for each type of test and sample. It will normally be sufficient to investigate the uncertainty only once for a specific method, and use the information to estimate the measurement uncertainty for all tests performed within the scope of that method.

Detailed information about measurement of uncertainty can be obtained in documents such as the EURACHEM/CITAC Quantifying Uncertainty in Analytical Measurement (Ellison & Williams, 2012).

9.6 Conclusions

It is not always easy to establish a QC policy and plan in an analytical laboratory for residues, despite the globally preconized norms for QC and

QA. Then it is very important to get a pragmatic view of the laboratory board in order to establish a feasible QC.

Sampling an agricultural raw material or product should result in a sample that preserves the original characteristics, so some care must be observed to guarantee this requirement. Chain of custody is the document for this guarantee.

A LIMS is a software tool dedicated to manage lab data, instruments, and workflows in order to improve lab productivity and efficiency. Then its implementation for residue analysis is very desirable for large sample quantities.

The critical evaluation of a result for its reporting and interpretation should be conducted taking into account several figures of merit as accuracy, linearity, LOD, LOQ, precision, selectivity, sensitivity, and robustness. The analytical report should present these figures in its content.

Finally, each measure has an uncertainty associated with it, resulting from errors originating from the various stages of sampling and analysis, and from imperfect knowledge of factors affecting the result. And a statement of the uncertainty associated with a result conveys to the customer the quality of the result.

References

Barwick, V. (Ed.) (2016). *Eurachem/CITAC guide: guide to quality in analytical chemistry: an aid to accreditation* (3rd ed.). ISBN 978-0-948926-32-7. <https://www.eurachem.org/images/stories/Guides/pdf/Eurachem_CITAC_QAC_2016_EN.pdf> Accessed 02.21.

Cooperation on International Traceability in Analytical Chemistry (2021). <https://www.citac.cc/> Accessed 02.21.

Currie, L. A. (1995). Recommendations in evaluation of analytical methods including detection and quantification capabilities. *Pure and Applied Chemistry, 67*, 1699−1723.

Ellison, S. L.R. & Williams, A. (Eds) (2012). *Eurachem/CITAC guide: quantifying uncertainty in analytical measurement* (3rd ed.), ISBN 978-0-948926-30-3. <https://www.eurachem.org/index.php/publications/guides/quam> Accessed 02.21.

Eurachem (2021). Eurachem—a quick reference. <https://www.eurachem.org/index.php/mnu-about> Accessed 02.21.

Olivieri, A. C. (2015).) Practical guidelines for reporting results in single- and multi-component analytical calibration: A tutorial. *Analytica Chimica Acta, 868*, 10−22.

Thermo Fisher Scientific (2021). LIMS—Laboratory information management systems. <https://www.thermofisher.com/br/en/home/digital-solutions/lab-informatics/lab-information-management-systems-lims.html?ef_id = EAIaIQobChMInf7z5uzf7gIVCwuRCh35CwwXEAAYASAAEgIJ3PD_Bw-E:G:s&s_kwcid = AL!3652!3!495353477681!e!!g!!!lims&gclid = EAIaIQobChMInf7z5uzf7gIVCwuRCh35CwwXEAAYASAAEgIJ3PD_BwE> Accessed 02.21.

CHAPTER 10

General remarks and conclusions

Remarks and conclusions are approaches used in order to facilitate the book understanding and its applicability in real situations. Remarks cover the most representative statements described in the text when conclusions summarizing critically the conducted discussions.

Then both can be used as a practical compilation which take part of relevant content from each chapter to present technical and scientific subsidies for strategic planning and for laboratory work. Fig. 10.1 illustrates this approach, which comprises the reading of the text, the understanding of its content and the application of its fundamentals to residue analysis.

10.1 Remarks

These remarks comprise those main points explored in each chapter in order to highlight the content of each one.

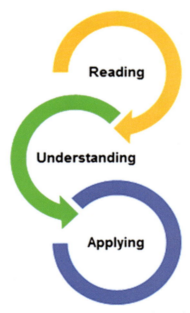

Figure 10.1 Steps involved in the use of the remarks and conclusion compilation. *Credit: author.*

We saw in Chapter 1, Introduction to Organic and Inorganic Residues in Agriculture, that agriculture is a large economic activity that comprises the production of raw material for food, feed, chemicals, pharmaceuticals, among other products. According the Food and Agriculture Organization of the United Nations (2018), the global agricultural production in 2016 achieved 4.7 thousand tons for the five main items produced (sugarcane, maize, wheat, rice, potatoes). For the livestock production, it achieved 27.6 thousand heads for the five main items (chickens, cattle, ducks, sheep, and goats). From this point of view, agricultural practices and food chains need chemical substances, as pesticides, fertilizers, additives, stabilizers, preservatives, antioxidants, antibiotics, sanitizers, and others. These chemical substances are known as agrochemicals and are used from tillage to harvest and processing steps. The production steps and uses generate a variety of sources of chemical residues that demand attention due to their negative impact on the environment and overall human health.

We saw in Chapter 2, Agricultural Matrices, that agricultural and related matrices comprise a large diversity of organic and inorganic materials for chemical analysis to determine presence or absence of chemical residues. Despite their heterogeneous compositions, these matrices can be studied by means of several classes of analytical techniques such as chromatography, spectroscopy, and electrochemistry. Environmental chemistry, as the branch of the chemical sciences that studies the processes involved with the dynamic of the chemical species (e.g., molecules and ions), can provide information to understand the fate and dynamics of these residues in the environment. On the other hand, food chemistry can provide information to understand also the fate and dynamics of these residues in the food supply and in humans. Both scientific branches can contribute to achieve reliable analytical results—that is, the correct interpretation of the generated analytical result. In this context, it is very important to define a sample—a part of a certain material to be analyzed—such as the matrix (the medium to be analyzed) plus the analyte (the chemical species of interest for the analysis). Then it is easier to establish the correct analytical approach, that is, the best methodology to be applied. Furthermore, sampling is the first experimental step to obtain the correct portion to be analyzed.

We saw in Chapter 3, Toxicology in Agriculture, that ecotoxicology and toxicology are both interdisciplinary scientific branches that study the effects of chemical substances on the environment (the first) and humans (the second) in order to define limits for the exposition to those chemicals—for example, pesticides and veterinary drugs. It involves chemistry, biochemistry, biology, pharmacy, and medicine being chemical analytic tools to access the necessary data to formulate and to confirm the required scientific hypothesis behind the established limits or concentrations for monitoring and control. Chemical analyzes—especially those conducted by means the use of chromatographic, spectroscopic, and electrochemical techniques—are paramount in obtaining data to estimate ecotoxicological and toxicological parameters (e.g., LD_{50} and LC_{50}) related to the presence of chemical residues from agriculture in several analytical matrixes (e.g., food, soil, water, among others).

We saw in Chapter 4, Fundamentals of Analytical Chemistry, that the role of analytical chemistry in the analysis and understanding of the chemical residues from agriculture effects on humans and on the environment is paramount. The presence and effect of a certain compound just are possible to be defined if analytical techniques and methods are applied, comprising identification, exposure assessment, risk characterization, and legal regulation aspects (Pierzynski, Sims, & Vance, 2005). Generally, chemical analysis can be considered as the use of concepts of analytical chemistry and its techniques and methods in the investigation and solution of real problems of variable complexity in different scientific or technological areas. The chemical analysis can generate information of both qualitative and quantitative character. In order to understand the application of analytical techniques for the analysis of agrochemical residues in several matrixes, it is of fundamental importance to introduce some basic terms of analytical chemistry, taking into account characterization, identification, and determination approaches.

We saw in Chapter 5, Main Analytical Techniques, that chemical analyses play an important role in agriculture and related areas, as supporting technologies at all stages of agroindustrial chains such as grains, forests, pulp and paper, and agricultural waste, among other sources of agricultural products. Furthermore, chemical analyses give the knowledge of chemical composition and presence or absence of contaminants in food and pollutants in the environment. Then, it ensures the quality and the reliability of agricultural products and processes from the producer to the consumer. Sampling is the initial step that ensures a sample is representative of the material from which it is taken, paying attention to the need to minimize errors and that the sample is composed of the matrix plus the analyte. The preparation is the stage in which the sample goes through a procedure that aims to make it physically available for separation and/or detection (e.g., grinding, solubilization, digestion, partition extraction), emphasizing that in some techniques, the prepared sample can proceed directly to the detection. In the separation step, the sample is divided into its chemical constituents from solute—solvent interaction mechanisms (e.g., physiosorption and chemisorption), while in the detection step, the intensity of the analyte response to the detector's operational principle (e.g., electric current, absorbed radiation, emitted radiation) which leads to an analytical result to be interpreted to generate the required information—this concept is typical for chromatographic techniques and their methods. The control and analytical monitoring of agrochemical residues usually requires chemical analyzes that can cover a large number of samples at a low cost (approach 1). On the other hand, more refined studies already require chemical analyzes of complexity and higher costs (approach 2). As the chapter's aim was to introduce the main analytical techniques for agrochemical residues in order to carry out analytical studies, we focused on those most representative classes of analytical instrumentation in order to generate quantitative data whenever possible, as spectroscopy, mass spectrometry, chromatography, electrochemistry, sensors, probes, and bioassays. Furthermore, special attention was dedicated to the extraction techniques.

We saw in Chapter 6, Analytical Methods to Selected Matrices, that analytical methods represent the functionality of the analytical techniques and the analytical chemistry fundamentals. They will use theoretical concepts (e.g., the absorption and emission of radiation) behind techniques considering figures of merit in order to construct the better approach to solve a certain problem, that is, the determination (quantitative or qualitative) of an analyte in a matrix. When we consider food and environment-related matrices it is paramount that methods should be robust[1] because the first are high-heterogeneous chemical mediums of analyzes. Moreover, multiresidue methods are preferable instead of one-single method. Additionally, special attention should be given to the use of certified reference materials to validate the method in order to guarantee its quality control (seen in the Chapter 4, Fundamentals of Analytical Chemistry). A reduction in the analysis steps, in energy and reactants consumption, and in the residue generation will provide a desirable green method.

---

[1] Robustness has been defined as being the capacity of an analytical procedure to produce unbiased results when small changes in the experimental conditions are made voluntarily. For more information: https://doi.org/10.1016/j.microc.2016.12.004

We saw in Chapter 7, Analytical Chemistry Towards a Sustainable Agrochemistry, that nowadays, agriculture is constantly required to become more sustainable, with reduction of its negative impacts on environment and public health allied to an increasing in their positive impacts on society and economy. These are challenges and, at same time, opportunities for new production systems for whose agrochemicals are indispensable. How could it not be different, these opportunities are extended to the use of emerging technologies and approaches to support a sustainable agrochemistry proposal in order to reduce negative impacts from agrochemicals to public health (e.g., occupational diseases) and to the environment (e.g., pollution of air, soil, and water). Analytical and environmental chemistry allied to technological chemistry input techniques, technologies, and knowledge to analyze, produce, and monitoring agrochemicals. Nanotechnology and biotechnology are new technological approaches to be incorporated to the agrochemicals for the best agronomic usages. Food security and environment are closely related to the agrochemical effects and near to the consumer perception, implying in laws, market restrictions, and public opinion. Finally, sustainability is a demand from society for greater quality of life and greater transparency in the productive chains.

We saw in Chapter 8, Sorption Study for Environmental Purpose, that the introduction of antimicrobial agents in therapeutic use for livestock and poultry was revolutionary in combating diseases caused by bacteria in the middle of the last century. In animal feed, it contributed to the considerable increase in animal production. However, this is currently under discussion because of the development of bacterial resistance and its impact on the treatment of diseases in humans. The use of antibiotics in animal production is considered by the World Health Organization an increasing risk for human health (World Health Organization, 2020). Then it is an example of a situation that demands the residue analyses in food and environmental matrices. The fate of agrochemicals in the environment and their effects are a subject of research for academic and regulation purposes. For instance, a certain agrochemical after its application can achieve soil components, be transported, and reach the surface water or groundwater, producing deleterious effects to humans and animals. These aspects can be evaluated by means the use of analytical techniques—especially chromatographic techniques—allied to mathematical models and treatments and considering molecular physicochemical processes.

We saw in Chapter 9, A Practical Guide for Residue Analysis that there are relevant difficulties to be overcome in order to reach the best conditions for the residue analysis. These issues comprising difficulties for the establishment of a feasible quality control strategy, problems in the sampling step, laboratory management, a correct reporting and interpreting analytical results, and the correct estimation of measurement uncertainty of results.

10.2 Conclusion

These conclusions were compiled from the content presented and discussed in each previous chapter, in order to reinforce to the reader those main points and statements in each ones.

Firstly, agriculture is a large economic activity that comprises the production of raw material for food, feed, chemicals, pharmaceuticals, among other products. From this point of view, agricultural practices and food chains need chemical substances (agrochemicals) to production and processing. However, these chemicals produce organic and inorganic residues whose need to be analyzed in order to ensure the food quality and to avoid environmental pollution.

Second, agricultural matrices are very heterogeneous regarding their origin and chemical composition. In order to facilitate their understanding, they can be divided into environmental, food and feed, meat and other animal products, beverages and fruit juices, and agroindustrial residues. The presence of chemical residues on these matrices is subject of control by means regulatory agencies in order to attend technical criteria for environmental and food safety. To applied the best analytical approaches is paramount the knowledge of the composition and physicochemical properties of these matrices. Then it is possible to obtain the reliable analytical results.

Third, ecotoxicological and toxicological scientific branches plays a central role in the understanding of exposure, fate and effects of agrochemicals and their residues for organisms and for the environment. Associated to their terms and usages, chemical analyzes are the tools to generate data and information to establish, for instance, exposure limits to certain chemical compound—observing parameters as LD_{50} and LC_{50} and risk assessment—and to monitor its presences in food, soil, water, among other analytical matrixes.

Fourth, the role of analytical chemistry in the analysis and understanding the effects of the presence of chemical residues from agriculture on humans and on the environment is beyond of question. To achieve this role, chemical analyzes should follow the figures of merit in order to guarantee the performance and reliability of a certain analytical method. Developing and validating are required steps to assure the correct application of the analytical method in the agrochemical residues analyses. On the other hand, mathematical methods, especially chemometrics, offer the basis for data treatment and interpretation. Furthermore, QA/QC and green chemistry are aspects whose should be considered to explore, in a sustainable and concise ways, the best results provided by analytical chemistry.

Fifth, analytical techniques play an important role in the analysis of agrochemical residues in food and in the environment, giving the knowledge of chemical composition and presence or absence of contaminants and pollutants. Nowadays, there are a large number of analytical techniques available to the laboratories according their necessities based on samples (matrices and analytes). For agricultural purposes we can highlight spectroscopic and spectrometric techniques (e.g., UV-vis, Fourier transform infrared, near infrared, atomic absorption spectrometry, and optical emission spectrometry); chromatographic techniques (liquid and gaseous phases) hyphenated to a large variety of detectors; thermal analysis and microscopy. Moreover, sensors and probes has been gaining more and more space due to its ease of handling and the speed in generating results. However, extraction step deserves a special attention in order to guarantee the reliability of the analytical results.

Sixth, analytical methods based-on instrumental techniques, especially chromatography and spectrometry, are very useful in the analysis of organic and inorganic residues from agrochemicals in food and environmental related matrices. These methods can be adapted to several matrices according the analyst necessity and its technical expertise, which turns them a versatile set of tools for quality control and monitoring. On the other hand, it is highly desirable that such methods follow the principles of green chemistry.

Seventh, sustainability is, undoubtedly, mandatory for modern agriculture. Especially when considering agrochemical inputs, a source of potential pollutants for the environment and contaminants for food, feed and beverages. And residue presence is a risk to be studied and evaluated by means a holistic vision in order to guarantee its reduction or the desirable absence. In order to reach the sustainability condition, emerging analytical technologies (e.g., nanosensors) can be used allied to automation and miniaturization approaches. Furthermore, artificial intelligence is a set of tools to be exploited to extract more analytical information.

Eight, sorption isotherm is not a simple tool for environmental risk assessment. However, it can supply valuable data regarding the agrochemical fate across a certain matrix. As demonstrated, frequently is necessary the fitting of mathematical plots and models in order to achieve the best information. From the characterization of two soils in the State of São Paulo with different levels of organic matter (OM), the sorption of oxytetracycline (OTC) antibiotic was verified in Brazilian soils. It is dependent on the content of OM present, presenting a potential risk to the environment, resulting from its strong sorption capacity to such soils, which makes it difficult to transport by leaching and, consequently, less supply of antibiotics in groundwater. These observations also provide subsidies for further assessment of the environmental risk of the antibiotic to microorganisms and soil activity. OTC sorption to both soils was in accordance with the Freundlich isotherm (L-type). Such isotherms showed a sorption of homogeneous characteristic for these soils. Finally, the initially proposed experimental strategy can be replicated to another agrochemical molecule in soils.

Ninth, it is not always easy to establish a quality control policy and plan in an analytical laboratory for residues, despite the globally preconized norms for QC and QA. Then it is very important get a pragmatic view of the laboratory board in order to establish a feasible quality control. Sampling an agricultural raw material or product should result in a sample that preserves the original characteristics, so some care must be observed to guarantee this requirement. Chain of custody is the document for this guarantee. A laboratory information management system is a software tool dedicated to manage lab data, instruments and workflows in order to improve lab productivity and efficiency. Then it is very desirable its implementation for residue analysis for large sample quantities. The critical evaluation of a result for its reporting and interpretation should be conducting taking into account several figures of merit as accuracy, linearity, LOD, LOQ, precision, selectivity, sensitivity, and robustness. The analytical report should present these figures in its content. Finally, each measure has an uncertainty associated with it, resulting from errors originating from the various stages of sampling and analysis, and from imperfect knowledge of factors affecting the result. And a statement of the uncertainty associated with a result conveys to the customer the quality of the result.

References

Food and Agriculture Organization of the United Nations. *Statistical pocketbook 2018*. (2018). <http://www.fao.org/3/CA1796EN/ca1796en.pdf> Accessed August 2020.

Pierzynski, G. M., Sims, J. T., & Vance, G. F. (2005). *Soil and environmental quality* (3th ed.). Boca Ranton.: CRC Taylor & Francis.

World Health Organization. *Antibiotic resistance*. (2020). <https://www.who.int/news-room/fact-sheets/detail/antibiotic-resistance#:~:text=Bacteria%2C%20not%20humans%20or%20animals,hospital%20stays%2C%20and%20increased%20mortality> Accessed February 2021.

Index

Note: Page numbers followed by "*f*" and "*t*" refer to figures and tables, respectively.

A

Absorption spectroscopy, 114
Acceptable daily intake (ADI), 55
Accuracy, 87
Acetonitrile, 225
Active ingredient (AI), 56
Acute toxicity, 56
Adaptation, 54
Adsorption, 56
Adverse effect, 56–57
Adverse event, 57
Agonist, 57
Agricultural matrices
   agroindustrial waste
      pig slurry, 36
      waste of beer fermentation, 36
   animal products, 30–32
   beverages, 33
   environmental matrices
      soil, 23–24
      water, 25–27
   feed, 27–30
   food, 27–30
   fruit juices, 33
   meat, 30–32
Agricultural Revolution, 195–196
Agrochemical, 1, 2*f*, 246
   analysis of, 247
   classes, 199–200
   commercialization, 202–205
   demand for sustainability, 197–199
   effects of, 217
   fate of, 217, 249
   pillars of sustainability, 201–202
   regulation, 202–205
   sorption soils, 221–234
   usages, undesirable effects from, 201
Agroindustrial waste
   pig slurry, 36
   waste of beer fermentation, 36

Albendazole (ABZ), 177–178
Ammonia, 198–199
Analytical chemistry
   in 21st century, 86
   artificial intelligence in, 207–210
   calibration, 92–94
      external standard, 92
      internal standard, 93–94
      standard addition, 92
   characterization, 85
   chemometrics, 100–103
   determination, 85
   figures of merit, 86–90
      accuracy, 87
      limit of detection, 87–88
      limit of quantification, 87–88
      linearity, 87
      precision, 88
      recovery, 90
      robustness, 89
      selectivity, 89
      sensitivity/sensibility, 89
   green analytical chemistry, 105–107
   identification, 85
   outlook of innovative analytical approaches, 210–213
   quality assurance, 103–105
   quality control, 103–105
   role of, 85, 250–251
   validation, 94–100
      accreditation of analytical laboratory for residue analysis, 99–100
      evaluation of uncertainty of results generated, 97–98
      interlaboratory comparisons, 97
      interlaboratory studies, 96–97
      repeatability, 98–99
      reproducibility, 98–99
      systematic evaluation of factors influencing results, 97

Index

Analytical technique/methods
  based on instrumental techniques, 193
  bioassays, 143–145
  chromatographic techniques, 131–137
    gas chromatography, 132–134
    liquid chromatography, 134–137
  electrochemical techniques, 138–140
    electrophoresis, 139–140
    potentiometry, 138
    voltammetry, 138–139
  inorganic residues
    in environment, 182–186
    in juice, 186–190
    in meat, 191–193
  mass spectrometry, 129–131
  organic residues
    in environment, 171–173
    fruits, 174–177
    in meat, 177–178
    in processed product, 179–182
    in vegetables, 174–177
    wastes, 174–177
  probes, 140–143
  sample preparation, 150–167
  sampling, 146–150
    environmental samples, 146–149
    food samples, 149–150
  sensors, 140–143
  spectroscopic technique, 114–129
    atomic absorption spectrometry, 122
    atomic emission spectrometry, 123–125
    fluorescence, 117–118
    infrared molecular spectroscopy, 119–121
    molecular spectrophotometry, 114–116
    nuclear magnetic resonance, 126–129
    optical emission spectrometry, 123–125
    X-ray emission spectrometry, 125–126
Animal feed, 27–30
Animal product, 30–32
Antagonism, 57
Aqueous solution, 219
Artificial intelligence (AI), 86, 207–210
Atmospheric pressure chemical ionization (APCI), 129–130
Atomic absorption spectrometry (AAS), 122, 123f
Atomic emission spectrometry (AES), 123–125

B
Bayesian network (BN), 210
Benzimidazole (BZs), 177–178
Bioaccumulation, 42
Bioassays, 143–145
Bioavailability, 42
Biochemical demand (BOD), 26
Bioconcentration factor (BCF), 49
Biodegradation, 9–10, 42
Biomarker, 42
Biomass, 5
Biosensor, 42
Biotechnology, 197
Biotransformation, 43

C
Cadmium (Cd), 191
Calibration, 92–94
  curve, 241
  external standard, 92
  internal standard, 93–94
  standard addition, 92
Capillary electrophoresis (CE), 139
Cation exchange capacity (CEC), 23
Central composite design (CCD), 102–103
Certified reference materials (CRMs), 95, 171
Chain of custody, 239, 252
Chemical ionization (CI), 129–130
Chemical oxygen demand (COD), 26
Chemical substance, 1
Chemisorption, 43
Chemometrics, 100–103
Chlorofluorocarbons (CFCs), 45
Chromatography, 246–248
Codex Alimentarius Commission (CAC), 13
Cohort study, 60
Cooperation on International Traceability in Analytical Chemistry (CITAC), 238

Crops, international control policies for residues in, 12–13
Cytochrome P450 (CYP), 61

D

Decision tree (DT), 206f, 210
Degradation processes, 9–10
Dimethylarsinic acid (DMA), 187–188
Dispersive solid phase extraction (d-SPE), 212–213
Dissolved oxygen (DO), 26
Distribution constant, 150

E

Ecology, 43
Ecosystem, 43
Ecotoxicology, 39–55, 246–247
Electrical conductivity (EC), 26
Electrochemistry, 246–248
Electrolyte, 139
Electrolytic conductivity detectors (ELCD), 172
Electron capture detectors (ECD), 172
Electropherogram, 139
Electrophoresis, 139–140
Electrospray ionization (ESI), 129–130
Emerging pollutants (EPs), 5
 photodegradation of, 10
Endocrine disrupter, 43
Energy dispersive spectrometer (EDXRF), 125
Environmental assessment (EA), 43
Environmental chemistry, 246
Environmental impact assessment (EIA), 44
Environmental impact statement (EIS), 43
Environmental matrices
 soil, 23–24
 water, 25–27
Environmental monitoring, 44
Eutrophication, 44
Exposure assessment, 63

F

Fast atom bombardment (FAB), 129–130
Feed, 27–30

Fenbendazole (FBZ), 177–178
Figures of merit, 86–90
 accuracy, 87
 limit of detection, 87–88
 limit of quantification, 87–88
 linearity, 87
 precision, 88
 recovery, 90
 robustness, 89
 selectivity, 89
 sensitivity/sensibility, 89
Fingerprint, 119
Fluorescence, 117–118
Food, 27–30
 additive, 63–64
 international control policies for residues in, 12–13
 safety and standardization, 13–16
 security, 197, 249
Food and Agriculture Organization (FAO), 3
Fruits, 174–177

G

Gas chromatography (GC), 131–134, 173f, 211
Gel-permeation chromatography (GPC), 151
Good laboratory practice (GLP), 45, 64
Green chemistry (GC), 105–106, 250–251
Greenhouse effect, 45
Greenhouse gases, 45
Green Revolution, 196

H

Hazard
 assessment, 45–46, 65
 evaluation, 46, 65
 identification, 65
Hierarchical cluster analysis (HCA), 100–101
High performance liquid chromatography (HPLC), 134–137, 187–188
 mobile phase for, 150
Hydrolysis, 9

I

Inductively coupled plasma-mass spectrometry (ICP-MS), 187–188

J

Juice, inorganic residue in, 186–190

K

k-nearest neighbor (KNN), 100–101

L

Laboratory information management system (LIMS), 240
Lethal dose (LD), 46
Limit of detection (LOD), 87–88, 112, 210, 240
Limit of quantification (LOQ), 87–88, 112, 210
Linear discriminant analysis (LDA), 100–101
Linearity, 87
Liquid chromatography (LC), 131, 134–137, 211
  HPLC, 134–137
Lowest-observed-adverse-effect level (LOAEL), 46, 51

M

Mass spectrometry (MS), 129–131, 247–248
Matrix assisted laser desorption ionization (MALDI), 129–130
Maximum residue limit (MRL), 14t, 66
Meat, 30–32
  chemical composition of, 32t
  inorganic residue in, 191–193
  organic residue in, 177–178
Metabolite, 11
Methodology, 221–225
Microorganism, 221–224
Modus operandi, 90, 174f, 176f, 179f, 183f, 187f, 191f, 193f
Monomethylarsonic acid (MMA), 187–188
Monte Carlo simulation, 48, 68

Multiple chemical sensitivity (MCS), 68
Mycotoxin, 68

N

Nanotechnology, 197
Nanotoxicology, 68
Nitrous acid, 9f
Nonylphenols (NPs), 211
  chemical structures of, 211f
  extraction techniques for, 212f
No-observed-adverse-effect level (NOAEL), 48, 51, 69
NP ethoxylates (NPEOs), 211
Nuclear magnetic resonance (NMR), 126–129

O

Occupational exposure limit (OEL), 69
Occupational hygiene, 69
Optical emission spectrometry (OES), 123–125
  advantages of, 123–125
  limitations, 125
Organic compound, 26
Organic farming, 199
Organic matter (OM), 23, 252
Organization for Economic Cooperation and Development (OECD), 45
Oxfendazole (OXF), 177–178
Oxibendazole (OXI), 177–178
Oxytetracycline (OTC), 221
  chromatographic method of, 227f
  literature for, 229
  load and speciation of, 224
  physicochemical properties of, 225t
  soil, 230f, 231f, 232f
  sorbed species, 234
  sorption of, 252
  structural properties of, 224

P

Partition coefficient, 49, 70
Pentachloronitrobenzene (PCNB), 172
Persistent organic pollutants (POPs), 5, 50, 71, 171
Photodegradation, 9–10, 50

Photolysis, 9
Plant hormone, 199
Polystirene divinylbenzene (PSDVB), 225
Polyunsaturated fatty acids (PUFA), 31t
Potentiometry, 138
Precision, 88
Predicted no-effect concentration (PNEC), 50
Principal component analysis (PCA), 100−101
Probes, 140−143

Q

Quality assurance (QA), 98−99, 103−105, 237, 250−251
Quality control (QC), 86, 98−99, 103−105, 237−238
  analyses, 190
  strategy, 249−250
Quality management, 238
Quantitative structure−activity relationship (QSAR), 51, 72, 77

R

Ready-to-drink (RTD), 187−188
Recovery, 90
Reference dose (RfD), 51, 72
Relative standard deviations (RSD), 212−213
Repeatability, 98−99
Reproducibility, 98−99
Residue analysis, practical guide for
  chain of custody, 239
  estimation of measurement uncertainty of results, 241−242
  globally preconized norms for, 242−243
  laboratory management, 240
  quality control, 237−238
  reporting and interpreting analytical results, 240−241
Risk analysis, 52
Risk assessment, 52, 73
Risk management, 53, 73
Robustness, 89, 248

S

Sample preparation, 150−167
Sampling, 146−150, 247−248
  environmental samples, 146−149
  food samples, 149−150
Screening, 53
  level, 53
  test, 53
Selectivity, 89
Sensitivity/sensibility, 89
Sensors, 140−143, 247−248
Soft independent modeling of class analogy (SIMCA), 100−101
Soil, 23−24
  characterization, 221, 223t, 226
Soil organic matter (SOM), 219
Solid-phase extraction (SPE), 131, 151−167, 211
Solvolysis, 9
Sorghum bicolor, 27−30
Sorption, 53
  constant, 54
  isotherms, 217−221
  soil characterization, 221, 223t, 226
  soils-agrochemical, 221−234
Spectrophotometry, 114
Spectroscopy, 246−248
Standard(ized) mortality ratio (SMR), 74
Standard operating procedures (SOP), 240
Structure−activity relationship (SAR), 54, 74
Support vector machines (SVM), 210
Sustainability, 213, 251
  demand for, 197−199
  pillars of, 201−202
Synergy, 74

T

Thiabendazole (TBZ), 177−178
Tolerable daily intake (TDI), 75
Tolerance, 54
Toxicity, 55, 75
Toxicology in agriculture
  dietary exposure, 78−79
  disorders and diseases associated to agrochemical residues, 78

Toxicology in agriculture (*Continued*)
  ecotoxicology, 39–55
  environmental exposure, 80–83
  occupational exposure, 79–80
Toxin, 76
Trimethylamine, 9*f*

U
Ultra-high-performance liquid
    chromatography (UHPLC), 135

V
Vegetables, 174–177

Volatile organic compounds (VOCs), 147
Voltammetry, 138–139

W
Wastes, 174–177
Water, 25–27
Wavelength dispersive spectrometer
    (WDXRF), 125

X
Xenobiotic, 55
X-ray emission spectrometry, 125–126

Printed in the United States
by Baker & Taylor Publisher Services